CFZ YEARBOOK

1997

edited by
Jonathan Downes

Typeset by Jonathan Downes,
Cover and Layout by SPiderKaT for CFZ Communications
Using Microsoft Word 2000, Microsoft , Publisher 2000, Adobe Photoshop CS.

Photographs © 2007 CFZ except where noted

First published in Great Britain by CFZ Press

CFZ Press
Myrtle Cottage
Woolsery
Bideford
North Devon
EX39 5QR

© CFZ MMVIII

ISBN: 978-1-905723-31-7

Contents

Introduction to the 2008 Edition

This was always my favourite Yearbook. And it is fitting, I think, that it is the last one to be completed under our reissue programme, because it is still my favourite now, 11 years after it was originally finished, and not just because I don't have any more laborious reformatting, and page restoring to carry out. I was always particularly fond of this volume, because more than any of the others, it encapsulated my entire original vision for the Centre for Fortean Zoology:

It has exciting field research into apparently paranormal phenomena in a far-flung corner of the world. It has academic excellence in the form of an encapsulation of the available data on one gloriously obscure species of mammal, from three different perspectives, all of them academically peerless. It has the sort of archival research that I have always admired; laboriously dissecting the anomalous mammal reports for over 40 years from one English county. And it has groundbreaking cryptozoological research which lays one spectacularly stubborn cryptozoological canard to rest once and for all. It has a range of literary styles ranging from the academic, to the whimsical, and from the hard-core scientific to the downright weird.

Sadly, several of the contributors are no longer with us: Joan Amos and Clin Keeling are particularly missed. I was very fond of both of them, and they leave gaping holes in my life. Other contributors have drifted away from the CFZ. I have not heard from Justin Boote, or Tom Anderson in years. But others: Darren Naish, Karl Shuker, Mike Playfair, and Neil Arnold are still CFZ stalwarts, and Muirhead and I are *still* working on Hong Kong related fortean projects to this day, and have *still* not published a book on the subject.

Like the 1996 and 1998 volumes this has been republished in facsimile form as far as possible, to retain something - warts and all - of the flavour of the original volume. No quasi-Stalinist revisionism in *our* publishing schedule!

1997 was a weird year for me, but rereading this Yearbook brought back a lot of fond memories. We have achieved a lot in the last ten years, and I wonder what the next decade has in store for us.

Jonathan Downes
Director, CFZ
North Devon,
May 30 2008

The CFZ Yearbook 1997

Edited by

Jonathan Downes

"Life is what happens to you when you're busy making other plans"

John Lennon

Also By Jonathan Downes:

Books

Take this Brother may it serve you well (1988)
El Grand Senor (1991) *with Kim Andrews*
Riding The Waves (1992) *with Kim Andrews*
Road Dreams (1993)
CFZ   Yearbook 1996 (1995) *Editor*
The Smaller Mystery Carnivores of the Westcountry (1996)
The Owlman and Others (1997)

Records

The Mistake (1982)
Emotional Fascism (1984)
You took me up (1984)
Outside the Asylum (1990)
Pyramidiocy (1991)
Breakfast (1992)   *with Brian Storer*
Sexgodbaby (1992)
The Chicken sleeps Tonight (1993)
The Case (1995)
Contractual Obligations (1996)

INTRODUCTION

As Robert Burns would no doubt have said, 'the best laid plans of Animals & Men usually go wrong'. This book was practically complete when we advertised it in issue eleven of Animals & Men. All that remained to be done was to transfer what was already typeset and laid out on disk to hard copy for the printers.

But of course things aren't as simple as that in this fortean omniverse. The printers said that if we gave them the disk they could access the disk directly and produce a better quality product. "Oh Goody" said us. and gave them the disk which they promptly managed to erase. Had we made a back up copy? Of some of it ... yes. Of most of it ... a resounding NO! Then believe it or not, whilst we were drowning our sorrows at the thought of having to retype this book from scratch we left half of the hard copy we DID have in a pub!

This is why the 1997 Yearbook is a little late. Many apologies to everyone who ordered it early and has had to wait. We will try to be more efficient with the 1998 volume.

Particular apologies must go to Ben Chapman whose article was lost, unsalvageably, in cyberspace. It will be included either in a future issue of Animals & Men or in the 1998 Yearbook.

In this book there are articles covering a wide range of scientific and fortean disciplines. I have edited them when to my certain knowledge a piece was inaccurate, but neither I, nor the Centre for Fortean Zoology as a whole, take any responsibility for inaccurate statements made in any of these papers. Likewise, every attempt has been made not to infringe anyone's copyright and we have attempted not to print any item which could be seen as libellous. Any such item, however, is the responsibility of the individual writer, and not of the Editor or publishers.

In here we have something for everyone. The two main themes of this year's volume, however, are 'Cryptozoological Movies' and 'Cattle Mutilation/Chupacabras'. We have widened the base of contributors in this volume and therefore the range of ideas, some of which will seem heretical to the most hidebound 'fundamentalist' cryptozoologists, is also wider than last year.

If there is anything here that any reader disagrees with we guarantee an automatic right of reply EXCEPT where this would just propogate an annoying piece of in-fighting. We have seen too many other organisations destroyed, or at least sidetracked, by such internicene squabbling and we shall not allow it to happen here!

The function of The Centre for Fortean Zoology is not to 'spout' any 'official party line'. we are a forum for people to discuss a wide range of beliefs and scientific methodology, and the only belief which we propound is that surrealchemy is just as important as genetics, and that the testing of mitochondrial DNA is as essential as UFOlogy. As a wise man said to my last night on a different subject altogether. THERE ARE NO RULES.

I want to dedicate this volume to everyone who has helped the CFZ, and me in particular get through what has been a very difficult year. You all know who you are and so I will not even attempt to list you. Apart from you, I want to dedicate this book to the memory of Jessica Mitford, a great lady, a fine journalist and an inspiration to us all. (Incidentally, as I have noted elsewhere, she is probably the only member of her family that I HAVE NOT written a song about!) Also this volume is dedicated to another cultural icon who dies during 1996: Richard Berry - the man responsible for the greatest single work of art of the 20th Century (he wrote *Louie Louie*).

I hope that you enjoy this volume and find something within the procession of the damned, which follows over nearly two hundred pages which will interest you and hopefully inspire you to further feats of cryptoinvestigative methodology.

Enjoy.

JD

HIGH STRANGENESS: CATTLE MUTILATION IN MEXICO
A Preliminary Report

by

Dr. Rafael A Lara-Palmeros

coordinator of research for the
Centre for the study of Paranormal Phenomena, (CEPP) Mexico.
Also the Mexican correspondent for the Centre for Fortean
Zoology.

The Republic of Mexico, as never before. is utterly confused and seriously alarmed by a series of bizarre phenomena which have manifested themselves since the month of February this year in a number of Mexican States.

INTRODUCTION.

Never before in the history of Mexico has there been such a disquieting and strange phenomenon as cattle mutilation. There have been isolated and aberrant reports but they were seldom taken into consideration. An example of this can be found in the 1993 reports of the death of thirty horses in the state of Guegerro. The genitals, tongue and viscera were removed from the animals and the blame was put squarely on cattle rustlers.

Another report suggests that thirty sheep were slain by a mandrill-like creature in the state of Hidalgo, and nearly a year ago the town of Itzamanalco in Mexico State, (on the slopes of Mount Ajusco), reported the loss of an unspecified number of sheep, chickens and cows killed by giant cats. Despite an investigation by the School of Veterinary Medicine and Zootechnology of the Universidad Nacional Autonoma de Mexico, these 'beasts' have remained unidentified.

The following chronicle illustrates the events which have shaken the country over the last few months:

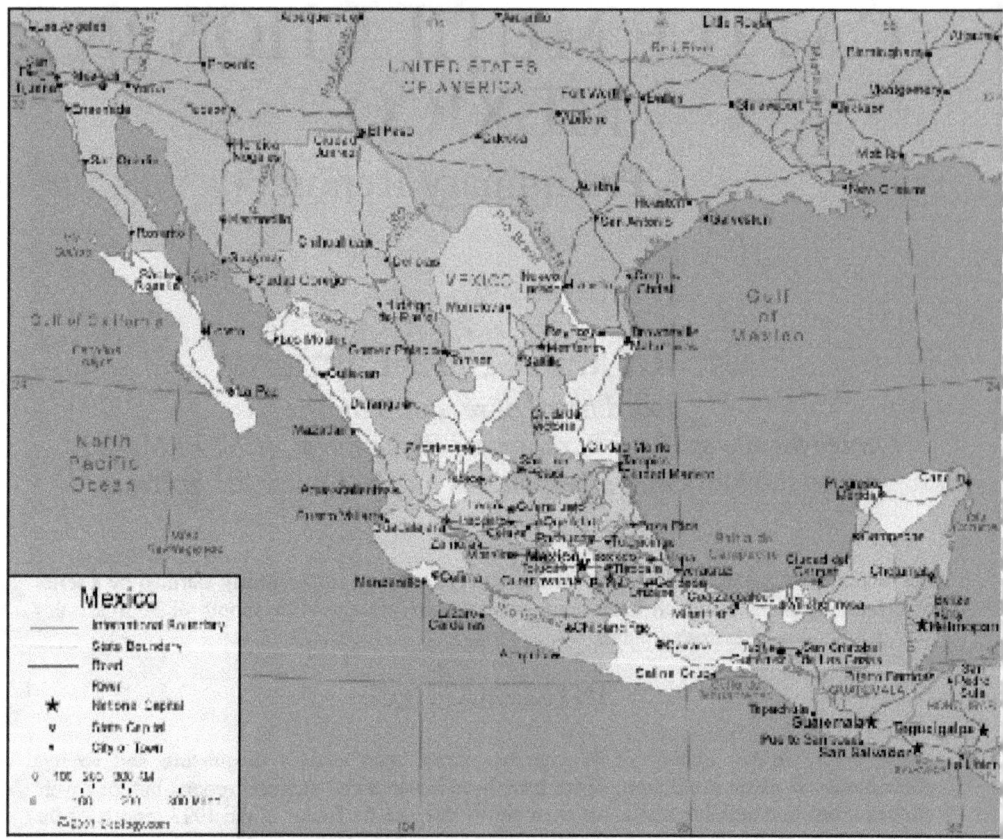

THE CASES.

On February 16th 1996, Professor Noe Montoya. a UFO researcher remarked in conversation to me that outlying communities of the city of Puebla were reporting strange deaths amongst their farm animals. These losses were reported to be blamed on a creature resembling the Puerto Rican Chupacabras.

I didn't believe this, however, on February 24th whilst participating in a television programme and discussing the important research being carried out in Puerto Ricoby Scott Corrales and Jorge Martin, I received confirmation that cattle mutilations were indeed taking place on the outskirts of Puebla.

I didn't really know what to make of this.

On March 4th 1996 Armando Gutierrez of Tampico, Tamaulipus State telephoned me. He mentioned the puzzling attacks on dozens of hens, four goats and fifteen chickens in the tiny village of Altamira, located about 2km from Tampico.

According to Dr. Gutirrez's testimony, the animals had wounds on their necks and had been left completely drained of blood. I was unable to confirm this by means of the press or other media which in itself made me doubtful.

It wasn't until March 29th, when Dr. Jesus Benitez, a fellow doctor at the Hospital General de Zona II, belonging to the Mexican Institute of Social Security informed me that 'Primer Impacto', a Television news programme, had mentioned animal mutilations from Florida as well as those from Mexico, particularly in jaliaco and Veracruz states.

The animals all had similar neck punctures.

The first official news item concerning an unknown predator roaming the Mexican countryside appeared on April 7th 1996. The location? Tlaxilixcoyan, Veracruz, where local cattle-men, municipal law enforcement and local authorities commenced a search for the enigmatic being and simaltaneously issuing a list of deaths attributable to it. Veterinarian Alfons Hernandez Perez expressed his belief that the deaths could not be attributable to either bats, dogs or felines.

On April 18th, Newscaster Jacabo Zabludowsky was openly mocking:

"....This being captured on the Internet, whose means of sucking remains a mystery, just another myth among the hundreds which surround us...."

His opinion on the subject would change dramatically two days later when he reported on the unexplained deaths in the community of Las Granjas, not far from Ciudad Guzman, Jaliaco. Over twenty sheep were found dead with the ubiquitous puncture marks on their bodies. They were completely drained of blood. Their owner, farmer Alberto Duran, could not believe what had happened.

On May 2nd, Victor Barragan, a veterinary of the Universidad de Guadalajara stated:

".... some feline and/or wild dogs were responsible for these massacres".

None the less, Teodosio Mendez Menza, owner of the Amecuaca Ranch, informed the authorities of the deaths of sixteen sheep, while at the same time, Bernadino Rodriguez of La Barranca de San Miguel experienced the loss of four sheep from their pen. Barragan remained unmoved, repeating that some natural cause, (pumas, feral dogs etc), was behind these incidents.

Meanwhile, Humberto Cota Gil, chief investigator ffor the Universidad Autonoma de Sinaloa declared that the 'Chupacabras' was nothing more than the Vampyrus spectrum, a bat species originating in Brazil, and whose size ranges between 50 and 70 cm. It sports 6cm fangs and a facial lobe of 25cm. Intermunicipal police patrolling the region of Guasava, Los Mochis and Bachomobambo were sent out to find this gigantic creature.

Vampyrus spectrum

The death list continued to rise in Tlalixocoyan, Veracruz. On May 3rd six dead sheep were found with marks on their necks and in the vertebral region. The settlements of Trancas, Palo Gancho, and El Nido were also affected by this spate of activity. According to a witness from El Nido, a dog like creature harrassed his flock, and when he attempted to shoot the intruder, it crossed a barbed wire fence without either wounding itself or making a noise.

In Sinaloa a brigade of specialists and policemen conducted a tour of inspection of all the affected municipalities as the death toll rose to a hundred animals lost. Jose Luis Garibay, a pathologist from Mexico's Department of Livestock and Agriculture determined that the animal must be enormous because it is quite capable of lifting 540kg, and slaying forty sheep simultaneously in order to consume twelve litres of blood from each of them. A sheep that had been left tied in a particular place at the Villa de Ruiz Cortinez was found some 200m away from where its owner had left it. The animal had been decapitated and exsanguinated.

Townspeople armed with rifles and machetes took to the countryside the following day in the hopes of finding the predator. According to the municipal secretary, Sergio Palencia Perez, no trace was found, although some claimed to have seen an enormous bat flying at a low altitude. In Los Mochis, Topolobambpo, the supernatural predator killed off forty sheep at Rancho La Remolacha, prompting a massive mobilisation of elements outfitted with infrared lights, shotguns, electrical equipment, helmets and riot shields.

This belligerent response did not keep the mysterious attacker from going about his appointed rounds. On May 5th, Tlalixcoyan awoke to discover twenty more sheep and twelve unspecified farm animals entirely bloodless. Based on the testimony of one Philogonio Jiminez, the Tlacoya Jimeno (Farming Co-operative) was shocked to discover thirteen sheep with wounds on their throats. Six turkeys were found dead in a similar fashion at Piedras Negras, and two of Rancho El Peladero's sheep were attacked by the nocturnal predator.

The village of La Loma, Jalisco. wasn't spared the blood letting. Ten of its sheep were slain by whatever roamed in the night while the dogs of its citizenry howled pathetically. It was on this strange evening that Jose Angel Pulido claimed to have been attacked by the 'Chupacabras'. He described the strange predator as a monster, over 80cms tall, and weighing some 30kg, with ashen-dark 'feathers' and sizeable wings.

EDITOR'S NOTE: There are a number of parallels between this sighting and indeed the other events recounted in this paper, and the notorious Owlman of Mawnan. At the risk of being acused of shamelessly plugging my book on the subject I would refer all interested readers to 'The Owlman and Others' (1997), and also to articles in A&M#6.

Juan Robles, a farmer from San Miguel Coahuila claimed that on the sixth of May 1996, four of his sheep were attacked by the alleged predator. This prompted engineer Angel Ramfrez to demand a serious formal investigation of the situation. This came about even as twelve lifeless goats were discovered Lagunillas de Huimilpan, Queretaro. They were exsanguinated and had the trademark perforations on their bodies.

At this point biologists began issuing their theories. Rogalio Sosa, a researcher at the Sinalca Centre of Science expressed the opinion that a serious investigation should be undertaken because the creature was obviously a strong one with formidable claws. His working hypothesis, however, was that the countryside was facing a 'rabid felid'. Luis Carlos Fierro, head of the Rural Livestock development dismissed the animal deaths as 'popular myths', assuring that the guilty parties were most likely coyotes, dogs, and cougars.

The 'Chupacabras' made an encore in the state of Veracruz this time at Comunidad Cruz del Milagro in Sayula da Alamen where it dispatched seven cows in its customary manner. Hundreds of miles away in the desert state of Chihuahua eight goats were attacked by an enigmatic beast, according to Ricardo Oropeza Medina of the Centre for Epidemiology of the state health agency.

On May 8th 1996 two sheep and six turkeys were slain in Tlalixocoyan, Veracruz. The 'manhunt' scheduled by the authorities did not take place since Atilano Martinez, the municipal president, and Armando Arguirre, the livestock farmer's representative chose to attend a political meeting rather than to comply with the wishes of the peasantry. The region was in the grip of tension and fear.

San Migle Coatlinchan, Texcoco, outside Mexico City made the news when it was found that eighteen birds were found dead and bloodless with marks on their necks. Residents believe that the numerous prints left on the ground by the animal are much larger than those left by any canid and suggest an animal with very large claws. In the wake of all these animal killings, livestock farmers began investing in large hunting dogs for help in their almost ceaseless patrols of their ranches. The president of the Regional Cattleman's Association admitted that some deaths had been caused by strange animal attacks. Zootechnician David Avila Figueroa of the University of Gudalajara had a contrary opinion. He claimed that the deaths were caused by a felid who was neither mysterious or abnormal and that animals being as sensitive as they are die from heart attacks (?!).

Vigilantes in the state of Nayarit fared somewhat better than those elsewhere. Six bloodless birds prompted a mobilisation and a vigorous pursuit of the creature. According to Mr Silva Avila:

"I saw it (the creature) in profile at a distance of some ten metres and it stood motionless, I picked up a stone and threw it at the creature but I missed. The animal turned toward me and then moved away with hurried jumps, vaulting over a two metre high wall. It measures some 80 cm, has a small head, and short pointed ears, like those of a bat. Its eyes were brilliant red".

He also added that the creature was covered in black fur with legs that bent at the knees and with two small arms like those of a kangaroo protruding from the torso.

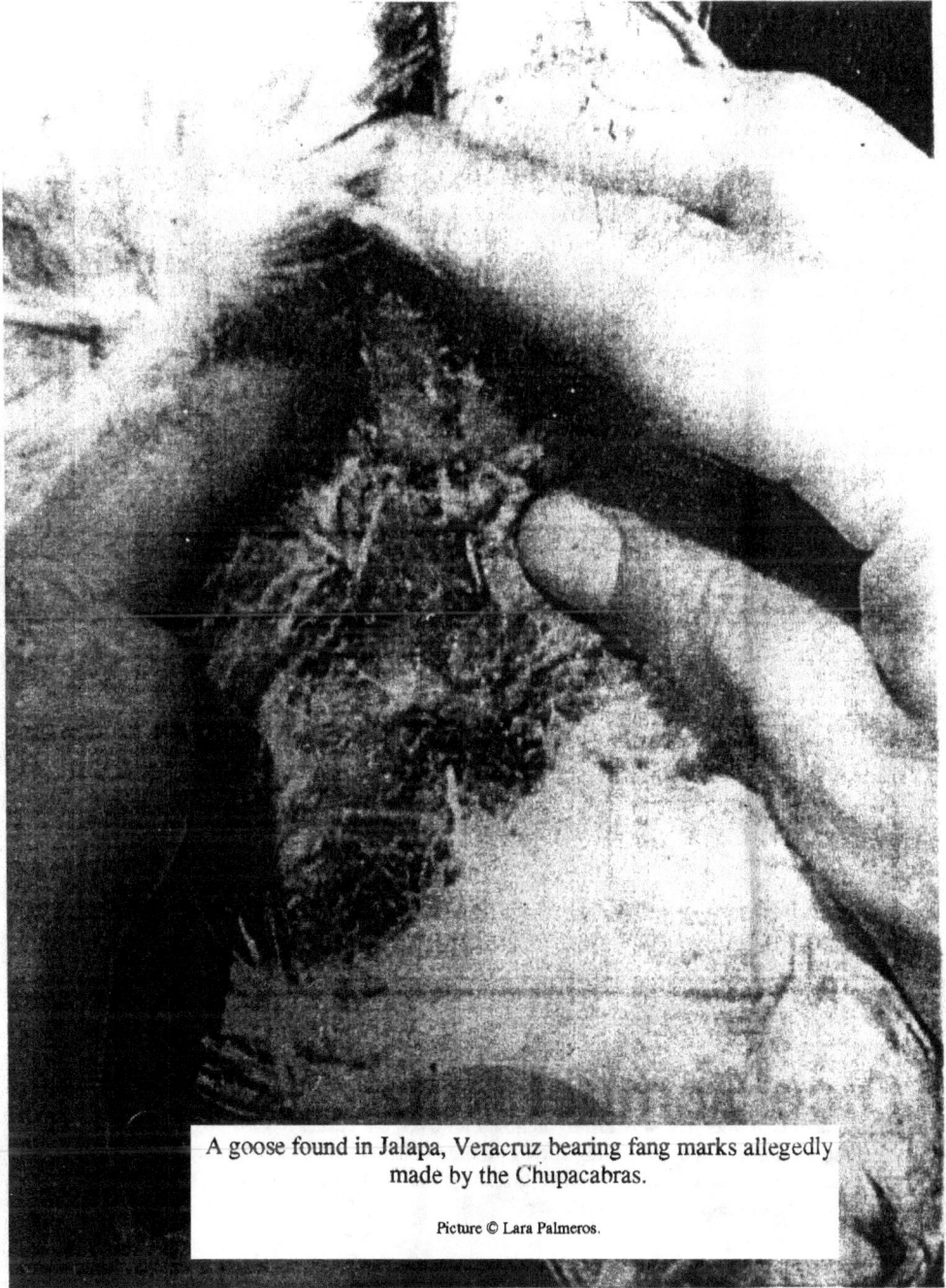

A goose found in Jalapa, Veracruz bearing fang marks allegedly made by the Chupacabras.

Picture © Lara Palmeros.

The creature returned to haunt the population once again causing the police to sally forth fully armed and ready to deploy tear gas against it. Meanwhile, in Cuatichan (Mexico State), nineteen birds and two pigs were found dead with the characteristic marks on their necks. The natives, curiously, began painting their houses and hanging crosses made of ocate upon their doorways..

On May 9th, the Archbishop of Morelia, Alberto Suarez Inda, begged the public to remain serene in the face of the phenomenon which was sweeping the country. He publically stated his confodence in the authorities. He declared that there exist:

"Very strange phenomena which are complicated to investigate and demand all the available time and means to be solved".

Antonio Camacho Diaz, a bishop of the Orthodox Catholic Church, suggested that the creatures could be a result of some congenital deformity or the result of radiation.

According to Carlos Vega Pineda, assistant director of municipal services, the so-called 'Chupacabras' is merely a misidentification of the well known *"Murcielago de las tumbas"* (the grabeyard bat) which has merely changed its dietary habits (???). As the mutilation epidemic spread special guards were established in the Municipalities of Texcoco and Zumpango, equipped with the added protection of crosses designed to scare the creature away.

Much in the same manner that Canovenas in Puerto Rico had become the favoured feeding ground of the 'Chupacabras', its Mexican equivelant appearred to have claimed Tlaxicoyan in Veracruz for itself. The citizenry, enraged and frightened by the depredations demanded that the authoritied do something to protect their animals and themselves, since the possibility that the creature might turn on humans was not taken lightly.

In Los Mochis, Sinaloa, one Benigno Cano claimed to have seen a strange creature some 90cms in height, with large, sharp fangs, red eyes, scales and a dragonlike appearance. This creature allegedly caused the deaths of two ducks, a goat and several chickens. Workers at the Abelardo L.Rodriguez dam stumbled upon the body of a dead bat measuring 85 cms in height and almost a metre wide with gigantic fangs. Mr Pavon Reyes, head of the Sonora Special Operations Group (GCES), said that he would concede that the *"Ocurrio Asi"* television programme could have the first bid on the story because he expected $50,000 for his trouble. It was later discovered that the creature was merely a dead cat which had been split open to make it appear larger.

Leon Guanahuato, a locale famous for its UFO related incidents was not spared a visitation from whatever it was that was killing animals all over the country. Two boys, Jesus Barajas and Julio Bermidez reported having seen a strange being. It measured approximately 1 metre and 20 cm, it was black skinned and had bulging eyes. They later described how it had cleared a two metre high fence, flying into the air.

On May 12th a veterinarian claimed that autopsies performed on dead geese suggested that the deaths could have been caused by a giant bat which might have been created by genetic manipulation either in the United States or in Mexico. This line of thought was echoed the following day by researchers from the School of Veterinary Medicine and Zootechnology at the Universidad Nacional Autonoma de Mexico. They thought that it was likely that mutations in canids or felids could have been the cause of the creature responsible for the killings that were now taking place all over the country.

On May 17th the residents of Ciudad Ayala in the state of Morelos had a repeat visit from the paranormal predator who struck at their flocks of sheep. The losses were attributed to a "giant vulture".

That same day in Tecomate, Sinaloa, frightened peasants used their machetes to kill an otter that they found in an irrigation ditch, believing that it was the 'Chupacabras'.

The same homespun 'experts' that had amused the Puerto Rican press turned up in Mexico. One of them believed that the 'Chupacabras' had been brought into the country from Africa. He also claimed that it was being utilised by the 'Golden Lucifer' sect for their rituals.

This being was allegedly able to sprinkle coloured powders at his vitims, thus putting them to sleep, and later attacking them. According to one 'expert' the 'Chupacabras' had a life-span of between fifty and a hundred years.

HUMAN ATTACKS.

Aside from the fact that a number of attacks on humans were confirmed throughout Mexico, it is worth mentioning the fate suffered by Mrs Teodora Ayala Reyes on April 29th 1996, in what was one of the most shocking cases in the Mexican 'Chupacabras' wave. It took place, as mentioned earlier in the village of Alfonso Genero Calderon in Sinaloa.

According to the lady's own description the attacker was a very bizarre creature. It was some 60cm tall, it gave off a strong stench and had large wings. Mrs Ayala was attacked at night, and she had bruises on her neck, her back and her face. Her assailant disappeared into the night.

As reported elsewhere, Jose Angel Pulido, a resident of Tlajomulco de Zuniga, was accosted by an improbable creature that inflicted wounds - hematomas and deep scratches - on his arms. The images of these two victims of alleged 'Chupacabras' mayhem were broadcast on Mexican television causing nationwide consternation.

Two other attacks, such as those upon Juano Tizoc Montenegro, and Elvira Meza, were never confirmed, although they received a great deal of media coverage. Nonetheless, these last two witnesses, coincide in their descriptions of the attacker.

THE CHIAPAS CASE.

This specific case is worthy of mention because it occured in a state which is undergoing a high level of political conflict, and because certain 'pseudoinvestigators' attributed the events to the activities of the Zapatista Liberation Army (EZLN) in an attempt to distract popular attention from what is going on in Chiapas.

On April 1st 1996, Mrs Julietta Calderon awoke to find almost twenty dead, blood covered sheep on her property - a ranch named San Antonio de los Sauces. The dead animals presented ripped throats, which according to the official version proved that the attackers had been dogs. Curiously, neither the watchmen, nor the shepherd, nor the inhabitants of the ranch, were alerted to the activities of a wild pack of feral canids.

According to Jesus Espinoza Ramirez, a technician with the Ministry of Agriculture, there is no doubt that the killings were perpetrated by dogs. This hypothesis is disputed by the owner of the ranch. Victo Manuel Samoaya recalled seeing a person, half a metre tall, naked and albino like, hiding amongst the bushes:

"I thought it was a child", he stated, adding that he and a companion went after the strange creature without success. The President of the Asociacion de Ovnicultores de Chiapas, Ernesto Sanchez Yanini, pointed out that in the twenty years he has devoted to cattle farming, he had never known so many animals be killed in a single attack. However, he did not dismiss the notion that a pack of wild dogs may have been responsible.

THE OTHER SIDE OF THE PHENOMENON: THE PSYCHO-SOCIAL PHENOMENA.

Since its first appearance in the national media, the 'Chupacabras', has captured the attention of everyone, from musicians to intellectuals. Numerous references to the creature have appeared in Mexico's printed media presenting an entity which has been variously linked to Subcomandante Marcos of the EZLN, with the myth of 'La Llorona' and to the poverty and lack of education affecting the country. The real 'Chupacabras', so to speak - is the drought and the hunger affecting the country, along with the current political crisis, unemployment and lawlessness. Delirious minds have created their own myths and legends as a result of the penury which has engulfed Mexico as a nation.

It should be noted that the 'El Financiero' newspaper, in its May 12th 1996 edition, published an extensive review of the phenomenon, ascribing its cause to the oft mentioned Mexican crisis. It went as far as to point out that this phenomenon is a rumour which has invaded the country due to the low educational levels among the population, which needs to be better informed in order to sort out truth from fiction. Paco Ignacio Tabo, a noted Mexican novelist and essayist, Jose Luis Cusvas, an artist, and many others have echoed similar sentiments. The 'Chupacabras' is a rival figure to Marcos. It embodies former president Carlos Salinas de Gortari, it embodies a Mexico, poverty stricken, manipulated, extorted and deceived by its fellow countrymen.

Faced with this 'evidence', one is led to believe that all the goings on in this world, whichn relate to this bizzarre phenomenon, are mere fantasies.

Yet I wonder. Are these fine gentlemen aware of what happens in cyclical rhythms in the fields of England, France, Sweden, Brazil, the U.S and Canada, among others, where mutilated and exsanguinated animals are found by the dozens? What would they think about the Exmoor beast? The Gevaudan beast? The recent wave of cattle mutilations in New Mexico and the 1993 wave of similar occurences in Germany? It is one thing to discuss, philosophise and get together with cronies in coffee houses without ever dirtying the soles of your shoes by investigating the scene where the events took place. In order to have grounds to discuss the phenomenon:

WE MUST INVESTIGATE, VISIT THE LOCATIONS IN QUESTION, INTERVIEW WITNESSES, ANALYSE THE RESULTS GATHERED, and refrain from issuing forth false conclusions, which in my opinion MISLEAD THE PUBLIC.

THE RESEARCH.

Upon discovering that one of the 'hottest' sites in the country was Tlalixcoyan, a community locate some three hours drive to the north of Jalapa, the state capital of Veracruz, we commenced making preparations to visit the site. Nonetheless, bearing in mind that the School of Veterinary Studies and Zootechnology of the University of Veracruz should have already been aware of the situation, we paid a visit to its facilities for an interview with its Director, Mr Emelio Zilli (who can be reached by telephone at 0091 (29) 86-02-44) and with academician Jesus Morales.

Surprisingly, these 'authorities' professed having no knowledge of the affair and rather nervously told me that I was the first civilian to have approached them with an interest in this subject. To assuage their fears, I told them that I wasn't an agent from the Department of State, or from National Security, merely a medical specialist with an interest in the subject. Their response was:

"Well, we'll get in touch with you if we find anything...."

Faced with this attitude on their part, I tried to contact Mr Augustin Morales. president of the Tlalicoyan Cattle Farmers Association Headquarters, in Oiedras Negras. (Telephone 0091-

-296-700-88). He claimed not to be aware of anything in particular, but would be happy to contact me once information on the subject had been brought to his attention.

To this date no further word has been received from either agency.

We visitied Tlalixcoyan on May 1st 1996. The high level of apprehension amongst the population was clearly evident, and when we spoke to a number of farmers and cattlemen, a constant appeared to be that no sound was ever heard from the areas in which the mutilations took place. It was only on the following day that the bloodless, mutilated animals were discovered.

Mr Pedro Hernandez told us that he had seen:

"... a kind of large rabbit that hopped along until it vanished into the vegetation. It measured approximately 80cms and was dark in colour (....) the following day I found three completely bloodless goats and four chickens in a similar condition. I don't know what happened...".

We investigated prints in the surrounding area but did not find any traces. Possible organic remains were examined without anything significant being discovered.

At this point in time there was a possibility that an official investigation, which would include competent authorities as well as civilian researchers would be launched by officials. It never took place.

According to three local children, Balthazar Harrera, Ramiro Rodriguez and Juan Hernandez, three dead sheep were discovered on their farm on April 6th. The hapless animals had been relieved of their blood, and the boys were unable to find any animal prints belonging to a dog, a coyote or a similar creature.

They told me that they had said nothing to the authorities out of fear of reprisals. This was a very significant aspect that we were able to pick up on during our stay in Tlalixcoyan. The region is constantly in turmoil since it contains marijuana and poppy fields, secret airstrips, and is a choice location for drugs and drug traffickers. In spite of these facts, the locals were so worried by the encroaching phenomenon that they requested the assistance of the authorities. This help never arrived, since the authorities in question emitted verdicts from the safety of their desks without bothering to evaluate the real concerns of the rural population.

On May 4th, television channel 5 transmitted a report stating the Coatepec region (close to Jalapa) as well as Xico had been visited by the "Chupacabras", which had slain six hens, four rabbits and a goat. We visited the area the same day, interviewing veterinarians, farmers, members of the Cattlemens' Association, and even workers of the local health care center. The results were negative. No-one was able to give any information; for which reason, we cannot say for sure if what transpired in the area was true or false.

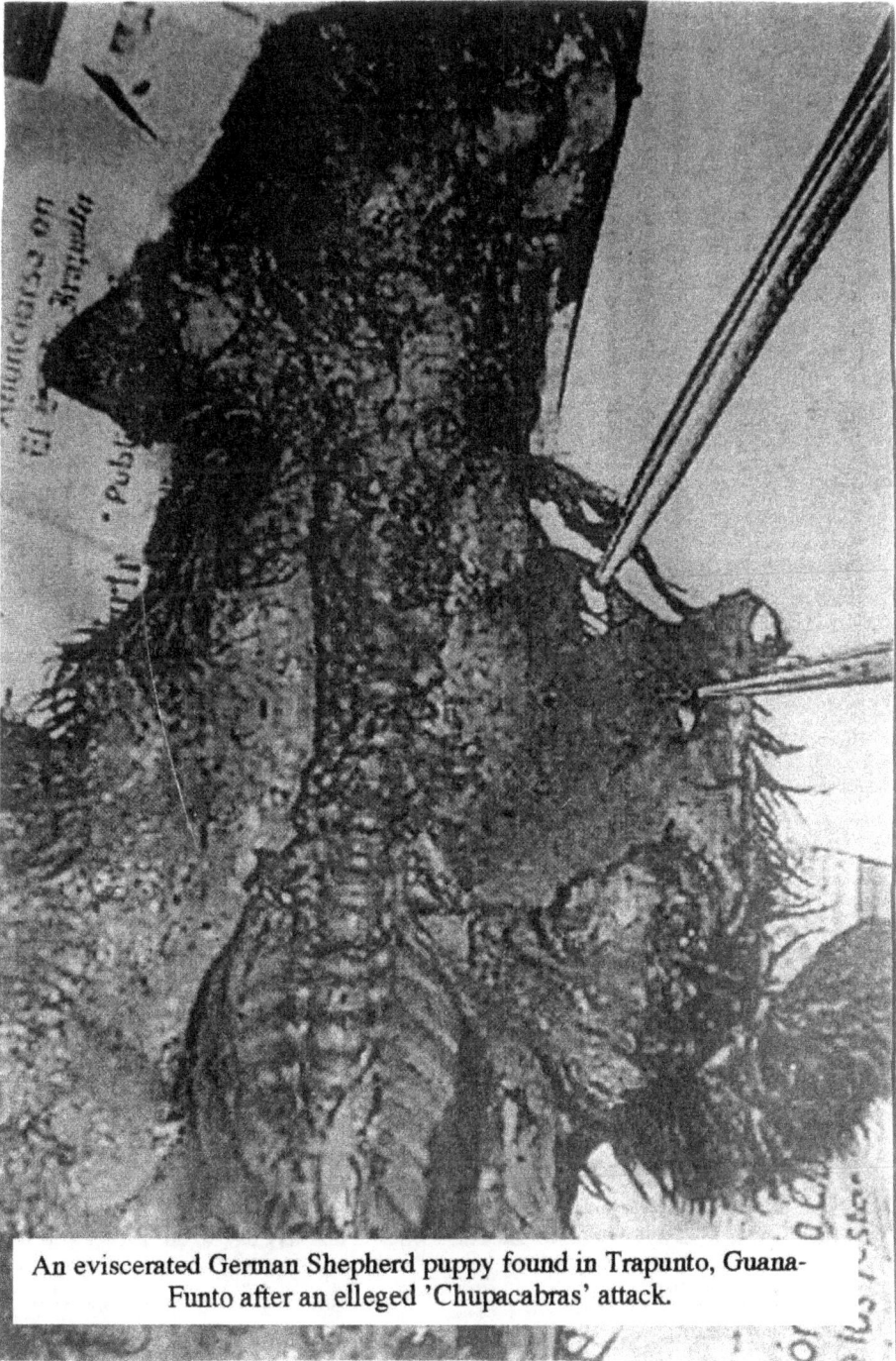

An eviscerated German Shepherd puppy found in Trapunto, Guana-Funto after an elleged 'Chupacabras' attack.

What conclusions can be drawn from the investigation performed? None. We can only consider that something or someone killed the animals, which were found with ripped throats or punctures on their bodies.

On May 11th, we investigated the case in Jalapa, visiting Colonia Independencia, where we were able to see the geese that had been slain and the puncture marks on their bodies. We interviewed Dr Villanueva, who commented that he had been unable to find any traces of blood on the birds. Residents and animal owners told us that they didn't hear anything at the time the killings took place.

It is also important to point out that some very strange reports have issued from the state of Jalisco, particularly from the Zapopan region, in which humanoid presences have been reported - women of a wolf-like appearance - in connection with the mutilation epidemic.

This has resulted in significant searches throughout the area without anything having been discovered to date. Witnesses insist that they have seen a woman with wolf-like characteristics, who had already attacked a horse and two people (as reported in "Cronica Policiaca" magazine, p19 #4, 7 May 1996). Curiously enough, this leads us to think about events which took place in the Dominican Republic during the 1970's involving mutilated animals and the presence of humanoid creatures.

On May 14th I placed a phone call to Dr Bernardo Villa in Mexico City. Dr Villa is an investigator with the Institute of Biology of the University of Mexico (UNAM) and is considered one of the world's foremost mastozoologists. During our conversation he remarked, "It is extremely important to conduct field research in the places which have been affected, following the scientific method: observation and exhaustive analysis of specimens. This should not be dismissed lightly, particularly with the psychosis that has gripped the population..." (As a sideline commentary we will add here that Dr Villa played a critical role during the investigation of the "Tecolutla Monster".)

ANALYSIS

As we stated in our introduction, Mexico has never faced as fascinating a phenomenon as the current one, and it is only logical to suppose that vague, silly, opportunistic, political and pseudo-scientific explanations should make the rounds, coupled with a thorough ignorance of the real phenomenon.

It is sad to notice the lack of a serious investigative criteria, backed by the scientific method, to face a situation of such proportions. It would seem that the explanations purveyed by different government bureaucrats are efforts at convincing the population that "nothing is going on in Mexico". The comments made by Juli Carabias, Secretary of the Environment, show the absence of a real awareness of the facts. The aim: to lead the population into believing that "it is all due to natural causes...", when to this very day no-one knows what is killing the goats, sheep, chickens, etc.

As an investigator of these events, I would take the following points into consideration:

1) The cattle mutilation phenomena is global, since it has occurred in France, Germany, Britain, the USA, Brazil, etc.

2) It is linked to a diversified phenomenology that includes the presence of large felines, winged entities, humanoid figures, etc (as in some of the cases presented here).

To date, there have been no UFO sightings liked to the phenomenon at the sites where killings have occurred (although they have been seen in others).

3) The phenomenon finishes as abruptly as it begins - folklorists and pseudo-philosophers attribute this to the tradition and collective unconscious of all peoples.

4) Real, objective information about the events must be made available to the population, free from bias and manipulation.

5) Scientific resources must be deployed in an effort to understand (if at all possible) what is causing these animal deaths.

6) Form brigades of veterinarians, zootechnicians, biologists, chemists, etc, charged with investigation each particular case openly and without prejudice.

7) The government should support the people involved in this research endeavour with police or even military vehicles if necessary.

8) Keep an open mind regarding the evidence.

It is indeed fascinating that, with the amount of research carried out, both from the bibliographical and field perspectives, the mystery still persists.

But this is Mexico, where the words of a misguided politician carry more weight than the statement of a true scientist.

Note: This is a preliminary report, "closed" on May 20th 1996. Nonetheless we continue to receive information from different states in Mexico.

REFERENCES:

*The following references are all newspapers unless stated
otherwise.*

1. "24 Hours" news program, 23 Jun 1993.
2. Pers Comm: Max Hofbauer, 10 Nov 1993.
3. Lara Palmeros, Rafael A *"Who are mutilated livestock in Iztlalamanalco (Mexican state)?"*
- Terra Incognita, CEFP's magazine, English version, issue 16, Mexico 1996.
4. TV program "Platicas de cafe" HGTV 4+ channel 24 Feb 1996: Jalapa, Eqz, Veracruz.
Mexico.
5. Pers Comm. Dr Armando Gutierrez. 14 Mar 1996.
6. Pers Comm. Dr Jesus Benitez. 29 Mar 1996.
7. "NotiVer" news program 1 Apr 1996. Veracruz, Ver.
8. "El Sol Veracruzano". April 7th 1996. Jalape, Ver.
9. "24 Hours" News Programme. April 18 1996. Mexico D.F.
10. "24 Hours" News Programme. April 20 1996. Mexico D.F.
11. "Diario de Jalapa". May 3rd 1996. Jalapa, Eqz. Ver.
12. "El Sol Veracruzano". May 3rd 1996. Jalapa, Eqz, Ver.
13. "Diario de Jalapa". May 4.1996. Jalapa, Ver.
14. "Sol Veracruzano" and "Sol de Los Mochis". May 4 1996. Jalapa, Ver. and Los Mochis,
Sinaloa.
15. "El Sol Veracruzano". May 5th 1996 Jalapa, Ver.
16. "El Occidental". May 5th 1996. Guadalajera, Jalisco.
17. "El Occidental". May 5th 1996. Guadalajera, Jalisco.
18. "El Grafico". May 6th 1996. Jalapa, Ver.
19. "El Sol de Sinalca" May 7th 1996. Culiacan, Sinaloa.
20. "El Sol de Chihuahua". May 7th 1996 Chihuahua, Ch.
21. "Diaro del Sur". May 7th 1996. Acayucan, Ver.
22. "El Universal". May 7th 1996. Mexico. D.P.
23. "El Sol Veracruzano".May 8. 1996 Mexico D.P.
24. "El Sol de Tampico". May 8 1996. Tampico, Tamaulipas.
25. "El Sol de Cuernivica. May 8th 1996. Jalapa, Eqz, Ver.
26. "El Sol Veracruzano".May 8th 1996 Mexico D.P.
27. "El Universal". May 9th 1996. Guadalajera, Jalisco.
28. "El Grafico". May 8th 1996. Jalapa Eqz, Ver.
29. "Diario de Jalapa". May 8th 1996. Jalapa, Eqz, Ver.
30. "Sol de Morelia". May 9th 1996. Morelia, Michescan.
31. "Sol de Toluca". May 9th 1996. Toluca. Edo de Mexico.
32. "Sol de Los Mochis". May 9th 1996. Los Mochis, Sinaloa.
33. "Sol de Tlaxcala" May 9 1996. Tlaxcala. Tlax.
34. "Sol de Mexico" May 9th 1996. Mexico D.F.

35. "El Sol Veracruzano". May 9th 1996. Jalapa, Eqz, Ver.
36. "El Grafico". May 9th 1996. Eqz. Ver.
37. "Diario de Jalapa". May 9th 1996. Jalapa, Eqz, Ver.
38. "Diario de Jalapa". May 9th 1996. Jalapa, Eqz, Ver.
39. "El Sol Veracruzano". May 10th 1996. Jalapa, Eqz. Ver.
40. "El Sol de Centro". May 10 1996. Jalapa, Eqz, Ver.
41. "El Mexicano" May 10th 1996. Ciudad Juarez, Chihuahua.
42. "El Heraldo de Chihuahua". May 10th. Chihuahua Ch.
43. "Diario de Jalapa". May 12th 1996. Jalapa, Eqz, Ver.
44., 45. "Diario de Jalapa" and "Sol Veracruzano" May 11th 1996. Jalapa, Eqz, Ver.
46. 47. "Sol de Morelia". and "Sol de Mexico". May 11th 1996. Morelia, Mich and Estada de Mexico.
48. "El Sol de Leon". May 11th 1996.Leon. Guanajuato.
49. "Sol Veracruzano". May 11th 1996. Jalapa, Eqz, Ver.
50. "Diario de Jalapa". May 11th 1996. Jalapa, Eqz, Ver.
51. "Diario de Jalapa". May 12th 1996. Jalapa, Eqz, Ver.
52. "Diario de Jalapa". May 13th 1996. Jalapa, Eqz, Ver.
53. 54. 55. "Diario de Jalapa", "Politico", "Grafico". May 13th 1996. Jalapa, Eqz, Ver.
56. "Grafico". May 13th 1996. Jalapa, Eqz, Ver.
57. "El Sol Veracruzano". May 13th 1996. Jalapa, Eqz, Ver.
58. "Diario de Queretaro". May 14th 1996. Queretaro, Qua.
59. "Sol de Toluca". May 14th 1996. Toluca. Edo de Mexico.
60. "El Occidental". May 14th 1996. Guadalajera, Jalisco.
61. "El Sol Veracruzano". May 16th 1996. Jalapa, Eqz, Ver.
62. "El Sol de Cuernavaca". May 17th 1996. Cuernavaco, Nor.
63. "El Sol de Sinaloa" May 17th 1996. Culiacan, Sinaloa.
64. "Diario de Jalapa". May 17th 1996. Jalapa, Eqz, Ver.
65. "El Sol Veracruzano". May 17th 1996. Jalapa, Eqz, Ver.
66. "El Sol Veracruzano". May 18th 1996. Jalapa, Eqz, Ver.
67. "Diario de Jalapa". May 18th 1996. Jalapa, Eqz, Ver.
68. "Grafica". May 18th 1996. Jalapa, Eqz. Ver.
69. "El Sol Veracruzano". May 19th 1996. Jalapa, Eqz, Ver.
70. "El Sol Veracruzano". May 20th 1996. Jalapa, Eqz, Ver.
71. "El Sol de Los Mochis". May 4th 1996. Los Mochis, Sinaloa.
72. "Alerta Ciudadana". Magazine. pp 2-3 #2 April 19th 1996. Mexico D.F.
73. "Alerta Ciudadana". Magazine. pp. 10-11. #5 May 10th 1996. Mexico D.F.
74. "Cronica Policiaca" Magazine p.19. #4. May 7th 1996.
75. "Cronica Policiaca" Magazine p.25. #5. May 13th 1996.
76. Obijo Leonte. "The Vampires of the Dominican Republic". Alien Contact Magazine, p.14-15. May 28 1980. Mexico. D.F.
    NOTE: Lagunillas's case was published in 'El Sol de San Juan del Rio'. May 6th 1996. San Juan del Rio. Queratero.

THE **C**RYPTOZOOLOGY FILES

by Mark North

THE CASE OF THE
CHUPACABRAS

DON'T BE PARANOID MY DEAR,
ITS ONLY A HARMLESS HEDGEHOG

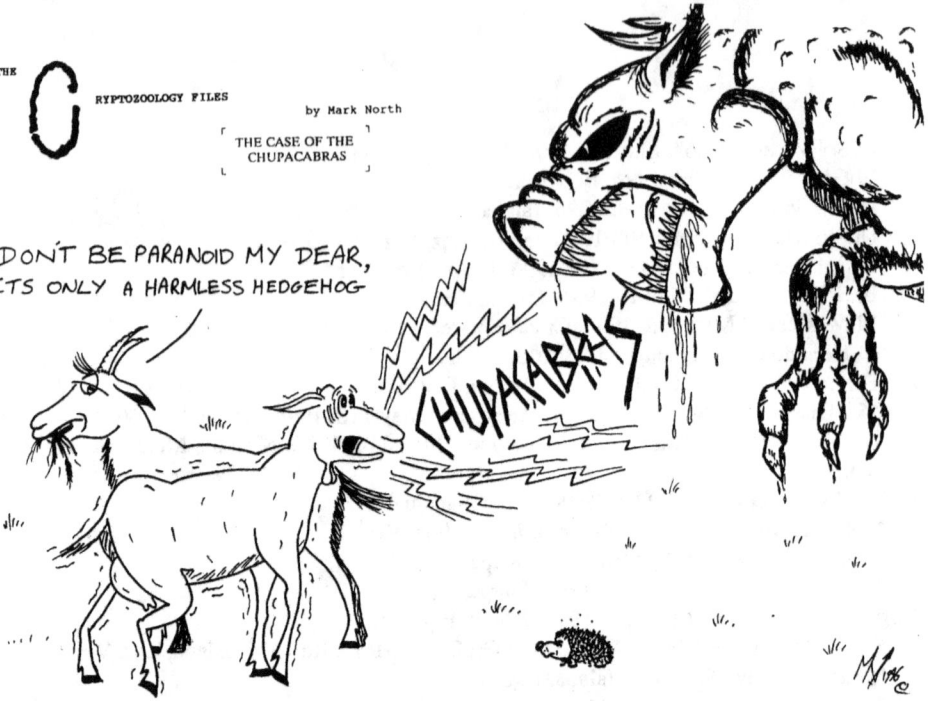

CHUPACABRAS

Am Fear Liath Mor : Ben MacDhui's Big Grey Man

by Dr Karl P. N. Shuker

This is the first time that this early Shuker article dating from the years before he was particularly well known within either fortean or cryptozoological circles, has been published in its entirety.

We are indebted to Dr. Shuker for letting us print it now, some ten years after it was originally written, with very few revisions.

When we consider the possibility of Great Britain housing large creatures still unknown to science, the most likely images conjured to mind must surely be ones portraying diffident inhabitants of certain Scottish lochs, or black pantheresque beasts stealthily prowling Exmoor's more remote terrain. Certainly very few people would contemplate a British abominable snowman! Nevertheless, in an attempt to elucidate scores of remarkable testimonies of visitors to at least one UK mountain, it has been postulated that this latter location may indeed shelter just such a creature.

Other opinions, conversely, reject this cryptozoological explanation in favour of theories incorporating an entity deriving from some twilight plane of existence bridging nature and supernature.

The mountain engendering such radical speculation is Ben MacDhui (Beinn MacDuibh in Gaelic), a towering peak culminating in a flat, barren summit. Indeed, at 4296 ft, not only is it the loftiest member of the Cairngorms, but it is also second only to Ben Nevis throughout Scotland. Nowadays, however, its principle claim to fame and the reason for its frequently used nickname of *"the haunted mountain"* is its mysterious occupant - referred to by the local Gaelic-speaking populace as Am Fear Liath Mor, or in English parlance, the Big Grey Man.

Not being the most readily accessible of mountains, Ben MacDhui has never been the focus of intensive and continuing exploration or scrutiny. It is usually reached via a 20-mile-long path veering away from the Glenmore Lodge and extending through the arduous but popular pass called the Lairig Ghru to Derry Lodge and Braemar, skirting Ben MacDhui's base at the north. The Big Grey Man, however, has added a further reason for the avoidance of this intriguing pinnacle by many climbers, and remains itself a totally inexplicable phenomenon.

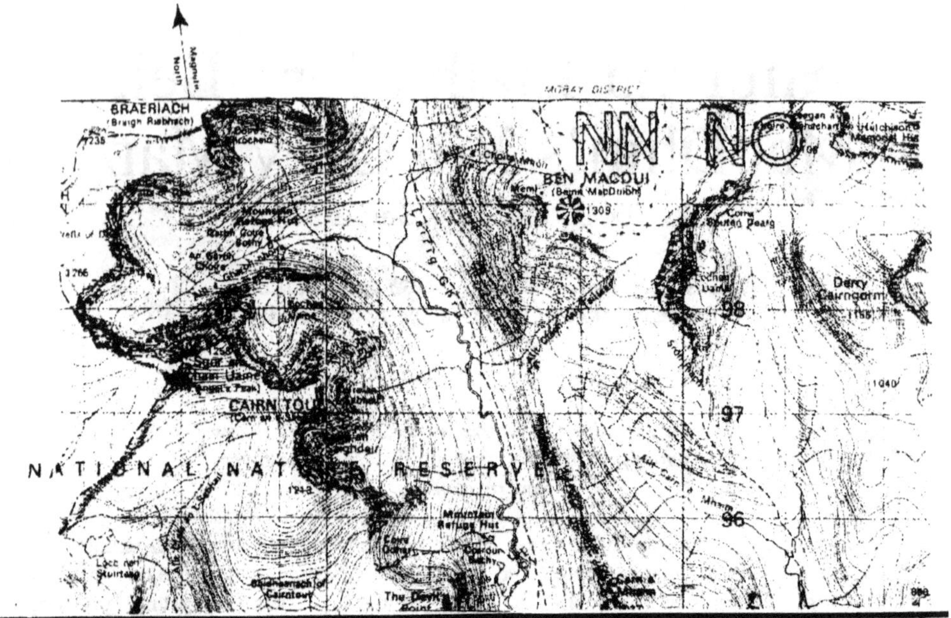
Magnetic North
MORAY DISTRICT
BRAERIACH
(Braigh Riabhach)
NN
NO
BEN MACDUI
(Beinn MacDuibh)
1309
Mountain Refuge Hut
Derry Cairngorm
98
97
96
CAIRN TOUL
NATIONAL NATURE RESERVE
Mountain Refuge Hut
The Devil's Point

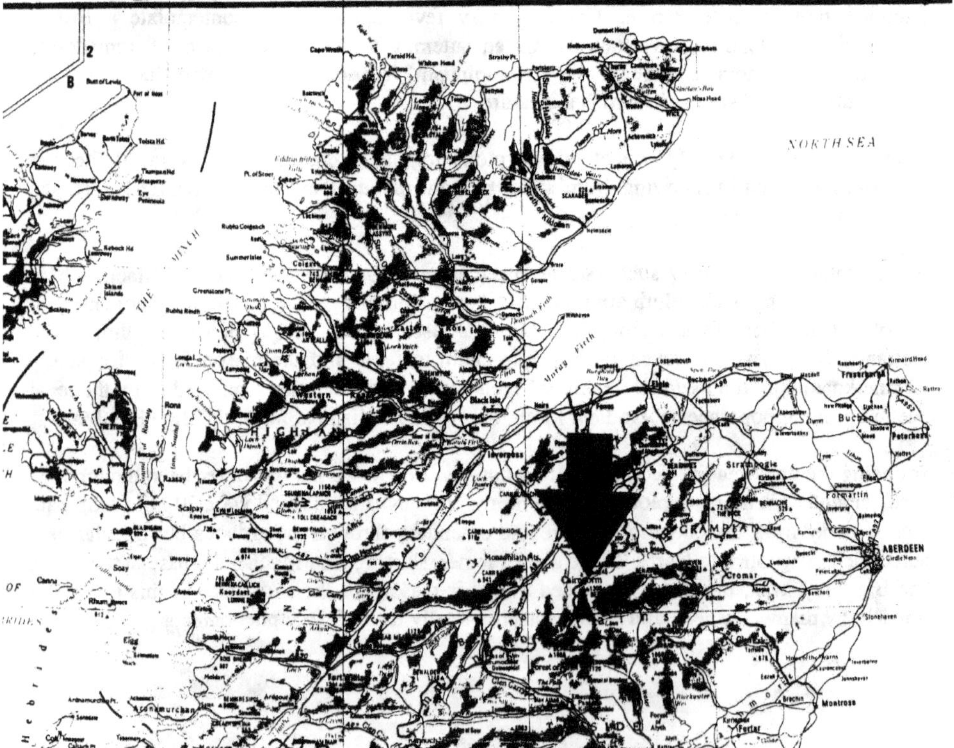
NORTH SEA
Cape Wrath
Butt of Lewis
Moray Firth
Inverness
Cairngorm
GRAMPIAN
ABERDEEN
Cromar
Buchan
Peterhead
Fraserburgh
Stonehaven
Montrose
Forfar
Brechin
Arbroath
Canna
Rhum
Eigg
Ardnamurchan
Hebrides

Unnerving incidents

A century ago, Am Fear Liath Mor was nothing more to the world-at-large than an obscure item of folklore from the immediate vicinity of Ben MacDhui.

However, as noted by *Enchanted Britain* author Marc Alexander, concomitant with the upsurging popularity of the relatively new sport of mountaineering, and the arrival of numerous enthusiastic climbers eager to scale Ben MacDhui's remote peak and traverse its shadowed wilderness, strange rumours and stories began to circulate of uncanny phenomena and unnerving incidents experienced by local and non-local climbers alike whilst upon this reclusive mountain. Nevertheless, such accounts were dismissed as imagination - until December 1925. For then, during the Annual General Meeting of the Cairngorm Club in Aberdeen (and even earlier in New Zealand), a singularly eminent and respected speaker recalled to a startled audience an anomalous event, which had happened to him whilst climbing Ben MacDhui in 1891. The speaker was none other than Norman Collie, an extremely proficient, internationally known mountaineer and also Professor of Organic Chemistry at the University of London, an austere man who shunned sensationalism. In short, someone who was certainly not given to wild imagination or fancy.

As subsequently documented fully in the *Cairngorm Club Journal* (July 1926), Professor Collie described how, immersed within a heavy mist, he had been descending from the Cairn at Ben MacDhui's summit when

"...I began to think I heard something else than merely the noise of my own footsteps. For every step I took I heard a crunch, and then another crunch, as if someone was walking after me but taking steps three or four times the length of my own."

At first he ridiculed himself for such fancies, but the sound persisted, though its agent remained concealed in the mist. As he continued walking

"...and the eerie crunch crunch sounded behind me, I was seized with terror and took to my heels, staggering blindly among the boulders for four or five miles nearly down to Rothiemurchus Forest."

He believed that there was:

"something very queer about the top of Ben MacDhui.."

and vowed not to return there alone.

Note well within his account (often distorted or embellished by subsequent writers) the crunching sounds resembling long striding footsteps following but not coinciding with his own, and his sudden uncontrollable fear and blind flight - in spite of his tremendous knowledge and understanding of every aspect of mountain climbing. For we shall meet these features many more times during this history of Am Fear Liath Mor.

A "big grey man"

Prior to the Professor's speech at Aberdeen, news of his intriguing experience had already reached another renowned mountaineer, Dr. A. M. Kellas, via the earlier but less-publicised New Zealand coverage. Consequently he wrote privately to Professor Collie regarding an incident that had occurred whilst climbing Ben MacDhui with his brother, Henry Kellas. Indeed, it was this incident that was responsible for providing Am Fear Liath Mor with its English appellation. For as related since in many publications, including R. Macdonald Robertson's More Highland Tales, Dr Kellas affirmed that he had actually spied a figure which he described as "a big grey man" walking out of the Lairig Ghru pass and around the 10-foot-tall cairn towards Ben MacDhui's summit, where it passed out of sight. He stated that it appeared to be at least as tall as the cairn itself!

Dr Kellas died in 1921 during the Mount Everest Reconnaissance Expedition, but an account related by his brother Henry to a close friend, Mr W. G. Robertson, was published in December 1925 in the *Aberdeen Press and Journal*. In this narrative, Robertson noted that the Kellas brothers awaited the being's reappearance at the summit:

"...but fear possessed them ere it did reach the top, and they fled. They were aware it was following them, and and tore down by Corie Etchachan to escape it".

Moreover, a third contemporary encounter involving a highly respected witness occurred in 1904, when Hugh Welsh and his brother, collecting biological specimens on Ben Macdhui for the University of Aberdeen, frequently heard the distinct sound of pacing footsteps, day and night; their origin was never discovered.

An immense newspaper-mediated interchange of opinions and comments followed the British publication of Professor Collie's account, and included many similar experiences and possible explanations. Indeed, a completely unexpected but extremely interesting perennial of the news media was born. Yet surprisingly, despite its potential, Am Fear Liath Mor has never attracted detailed investigation or scientific examination. Furthermore, until quite recently, even the numerous documented reports appertaining to this bizarre phenomenon remained widely dispersed among dozens of different publications, to the undoubted despair of any would-be researcher. Thankfully, however, a superlative collation of many such accounts by Cairngorms-bred author Affleck Gray, has since yielded his indispensible book entitled 'The Big Grey Man of Ben MacDhui' (1980), the definitive work on the subject.

Upon reading Mr. Gray's book. (which I heartily recommend to anyone interested in genuine mysteries), it became increasingly evident to me that the phenomenon of the Big Gray Man comprised a far more complex mystery than has hitherto been suggested.

Gigantic Footsteps.

The flood of correspondence following Professor Collie's revelation was stemmed by a relative lull of comparable events during the 1930's. However, the 1940's saw a remarkable resurgence of reports once more, many of which remain unparalleled even today in terms of significance. Hence the majority of accounts presented here have been selected from that period.

In the summer of 1940, as noted in *'New Highland Folktales'*, author R.Macdonald Robertson and a friend with whom he was residing on holiday spent a night inside Ben MacDhui's Shelter Stone (an enormous block of stone, beneath which travellers are indeed able to shelter and rest). Awakened by the growls of Robertson's bull terrier, the two friends clearly heard crunching footsteps approaching them along the gravel path, nearer and nearer. However, they never discovered their originator's identity, for the steps simply faded away, whereupon the terrier relaxed once more.

In her book *'The Secret of Spey'*, Wendy Wood describes an event that happened to her, also in 1940, upon the Lairig Ghru. While standing observing the striking scenery surrounding her, she heard what seemed to her to be an immensely resonant Gaelic-speaking voice.

Its suddenness and resonating timbre were so startling that she was rendered incapable of interpretation or even recollection of the words uttered. A few moments later, and seemingly beneath her, the voice rang out again.

At first, attempting to reconcile this uncanny occurrence with the possibility that someone was trapped beneath the snow and was calling for help, she searched the area meticulously, but all in vain. Stifling rising apprehension, she decided to leave the mountain. During her descent, however, she became aware that

"...gigantic footsteps followed me where no sound had been before..."

in long strides not coinciding with her own. Her hitherto suppressed fear now surfaced fully and she fled blindly downwards, directly exposed to the very same preternatural experience and accompanying fear that she had already been made aware of indirectly, by having read with interest Professor Collie's own encounter.

Strange figure

Affleck Gray records that a strange figure was sighted in 1942 by hardy climber Syd Scroggie from the Shelter Stone.

An impression of *Am Fear Liath Mor* by CFZ psychic
Phil Johnston.
© Phil Johnston/CFZ

Gazing into the night towards Loch Avon (A'an), he suddenly perceived

"...a tall stately human figure appear out of the blackness on one side of the loch and. clearly silhouetted against the water. pace with long deliberate steps across the combined burns just where they enter the loch."

The figure was not wearing a rucksack. and soon disappeared amidst the darkness upon the loch's opposite side. Scroggie rushed forthwith to the area and carefully searched for footprints or other signs of the figure's passing, but found nothing. He called out. but received no reply, experiencing instead an eerie sensation of unease, which spurred him to return with all speed to the Shelter Stone!

To my knowledge, 1943 saw the only instance on record of someone actually shooting the Big Grey Man! As he recalled in the Scots magazine (June 1958), mountaineer and naturalist Alexander Twenion had been walking along the Lairig Ghru in October 1943, hoping to shoot some game to supplement his wartime rations, when, whilst descending the track through the deep trough of Corrie Etchachan, he heard long striding footsteps behind him. He immediately recollected Professor Collie's narrative, peered through the mist. and was startled by a strange shape that loomed up, drew back. and then charged directly towards him! Instantly he drew out his revolver and fired three shots at the figure, but when it continued to approach him. unabated. he turned and fled downwards to Glen Derry.

An equally unnerving experience was faced by competent mountaineer Peter Densham whilst upon Ben MacDhui in 1945. Eating his lunch at the summit, he abruptly sensed another presence nearby - an impression frequently experienced by solitary climbers. Yet this time it was followed by crunching footsteps emanating from the Cairn on his left. Remembering the stories of Am Fear Liath Mor. he rose, still calm but anxious to investigate. Then. unaccountably, he was seized with apprehension and an intense desire to leave the mountain. He began to run. wildly, yet drawn irresistibly towards Lurcher's Crag (Creag an Leth-choin). Indeed, it was only by an immense effort of will that he was able to divert his path at the very last moment and thereby prevent himself from plummeting headlong over the cliff - the very one. moreover. over which. according to local legend. Am Fear Liath Mor, seeks to drive all visitors to Ben MacDhui!

In October 1965. the 'Weekly Scotsman' carried an article in which the writer related his own alleged encounter with a ten foot tall being whilst sitting at the Shelter Stone. surrounded by swirling snow and mist. Apparently the figure passed within five yards of the writer. and for once actually left behind visible footprints! These measured approximately fourteen inches in length, and were separated from one another by strides measuring four and a half to five feet! No photograph was taken. nor were casts made. Moreover. Affleck Gray failed to elicit any response to his letter sent to the writer concerned requesting additional details. Hence this potentially significant account cannot be followed up, and its veracity and implications remain unresolved.

Interestingly, as described in his book *'Romantic Speyside'*, author James Rennie has also encountered unusual tracks in this region of Scotland, though not upon Ben MacDhui itself but instead near Cromdale in Lower Speyside. Moreover, Rennie actually took photographs of these, which depict a long line of bilobed tracks comparable to the famous Devon Devil's hoofprints. Intriguingly, some of the local gamekeepers (already familiar with the spoor of known creatures of the region), to whom the photographs were later shown were totally perplexed. Others, however, appear totally ill-at-ease after having seen them. One affirmed that they are the tracks of a bodach - a form of Highland hobgoblin according to local legend. In actual fact, it is possible that these 'tracks' were the result of some meterological phenomenon.

Diverse.

From the sample of Big Grey Man reports presented here so far, three predominant, recurrent components can be readily identified :- a huge bipedal figure, long striding footsteps, and unreasonable fear experienced by climbers. However, as Affleck Gray documents in his book, many more accounts of mysterious happenings upon Ben MacDhui that have been attributed to Am Fear Liath Mor are also on record, and some of these involve a very more diverse range of features.

Strange Music.

One of Peter Densham's frequent mountaineering companions was Richard Frere, author of *'Thoughts of a Mountaineer'* and someone with comparable memories of inexplicable incidents faced whilst upon Ben McDhui, including one that he recounted in *'Open Air'* (winter 1948). Walking through Rothiemurchus Forest whilst contemplating the big grey man legends, he reached Laitrig Ghru and sat, deep in thought, staring at Lurcher's Crag. Without warning, he was seized by a spasm of unutterably deep, soul-draining despair and experienced the sensation similarly noted by Peter Densham of a nearby presence - very real, yet totally indefinable.

In addition, he also became aware of an extremely high singing note, which he soon determined via simple tests to be due not to any effect of reduced pressure impinging upon his eardrums. Throughout his ascent to the mountain's summit and subsequent descent to Lairig Ghru again, this surreal, ethereal music persisted, as did the sensation of an accompanying presence. Then, instantaneously, music and presence were gone, replaced momentarily by a flash of pure terror, then in turn by a tranquil, blissful peace as he re-entered Rothiemurchus Forest.

Further records exist of strange music, singing, even laughter, heard by climbers upon Ben MacDhui, regardless of any attendant presence, documented by Affleck Gray, who has actually experienced such originless strains himself, as he relates within his book.

So too had Hugh Welsh, who admitted to Mr. Gray, that even though he had proven categorically that many such instances which he had encountered had been due to a natural combination of running water and wind sighing through porous rocks, on certain occasions he had heard such music in the absence of any nearby water or wind, and had come away totally nonplussed.

Similarly, in a letter to 'The Scotsman' (September 2nd 1963), Seton Gordon recalled an incident in which he had been climbing Ben MacDhui with a friend in 1926, when his friend claimed that he could hear pipe music - even though the nearest house was many miles away!

In addition to these aural anomalies, certain sightings reported from Ben MacDhui are equally bizarre. Take the following accounts for example.

Great brown creature

The first was given to Richard Frere by an unnamed friend. To win a wager requiring him to spend a night alone upon Ben MacDhui, Frere's friends had pitched a tent beside the Cairn. However, after having experienced (while sitting outside) an explicable, disturbing deviation from his normally coolly logical mental state towards morbid self-analysis and despair, he retreated to his tent and fell asleep briefly. He awoke to perceive a brown shape standing between his tent and the moon. When the shape moved away, he immediately peered outside and, according to Frere, saw that apparently 20 yards away

"...a great brown creature was swaggering down the hill. He uses the word 'swaggering' because the creature had an air of insolent strength about it; and because it rolled slightly from side to side, taking huge measured steps. It looked as though it was covered with shortish brown hair... its head was disproportionately large, its neck very thick and powerful. By the extreme width of its shoulders compared to the relative slimness of its hips, he concluded its sex to be male. No, it did not resemble an ape: its hairy arms, though long, were not unduly so, its carriage was extremely erect."

Moreover, by applying trigonometry in conjunction with objects housed within his immediate surroundings, Frere's friend calculated that the creature was at least 20 foot tall. Thus, while agreeing in outline and striding footsteps with earlier reports, its height and especially its hirsute nature and apparent flesh-and-blood existence place it apart from any other Ben MacDhui report.

Another singular sighting involved a creature (?) seemingly more akin to the Cornish Owlman than to the Big Grey Man! Alfter turning around to locate the origin of footsteps suddenly sounding behind him whilst descending Braeriach (a neighbouring peak to Ben MacDhui), climber Tom Crowley spied a huge *"undefined misty figure with pointed ears, long legs, and feet with talons which appeared to be more like fingers than toes."* Interestingly, as Affleck

Gray notes. this is seemingly the only record of a mysterious giant entity from any other Cairngorm peak. Comparable beings have been reported. however. from certain other British mountains. A notable Welsh equivalent of the Big Grey Man is the Brenin Llwyd, also referred to as the Monarch of the Mist or the Grey King, and dealt with in Marie Trevelyan's *Folk-Lore and Folk-Stories of Wales* (1909). As documented in the excellent *Alien Animals* (1980) by noted researchers of modern-day mysteries Janet and Colin Bord, the Brenin Llwyd was a dreaded stealer of humans amidst the northern Welsh mountains, according to legend. Similarly, the Faroe Islands housed a corresponding version known as the Gryla, a comparably hirsute colossus, which inhabited their remote secluded peaks.

Returning to Ben MacDhui, the most macabre encounter of all, however, involves the entity that allegedly pursued authoress Joan Grant whilst upon the mountain in 1928. She describes it as being:

"Something utterly malign, four legged, and yet obscenely human, invisible and yet solid enough for me to hear the pounding of its hooves".

A Cairngorm centaur?!

To make matters even more perplexing, reports also exist concerning noticeable benevolent entities upon Ben MacDhui - very unlike the forbidding, malign being normally described by climbers. Hence, even the Big Grey man's hitherto omnipresent aura of malevolence is now in a state of flux.

Suddenly the uniform, well-defined (albeit still unexplained) phenomenon of An Fear Liath Mor is considerably diverse and markedly diffuse, to say the least. Added to this, the various explanations offered over the years as potential solutions to this bewildering mystery are themselves equally varied and comparably paradoxical. Indeed, the range encompasses cryptozoological, geological, psychical, psychological, climatic, mystical and ufological theories.

British Yetis.

A cryptozoological identity for the Big Grey Man is suggested primarily by the shaggy giant sighted by Richard Frere's friend, although on first sight its twenty foot height ostensibly destroys any notion of Ben MacDhui harbouring a race of British yetis or cognate crypto primates (e.g. American bigfoot, Chinese yeren, African Mulahu). For most accounts of these yield reported heights far less than this. Yet, as Dr. Bernard Heuvelmans relates in his classic work *'On the Track of Unknown Animals'* \(1958), evidence also exists for a 'super yeti' in Tibet - the giant nyalmo.

Even so, while all of the mystery primates listed above are superficially ape-like, the totally dissimilar build of the Ben MacDhui creature discussed here implies a quite different cryptozoological identity - one previously unconsidered moreover, relative to An Fear Liath Mor.

Some cryptozoologists feel that European tales of wildmen (actually given the formal scientific name of *Homo ferus* by Linnaeus), may have been based, not merely upon feral children and eccentric irascible hermits but instead upon historically recent encounters with a hominid type 'officially' extinct for thousands of years - Neanderthal Man. Moreover, relict populations of a Neanderthal type being known as the almas are apparently still living in remote mountainous regions of Mongolia, and during the early 1980's these were the subject of detailed field and bibliographical investigations by Leicester University archaeologist Dr. Myra Shackley.

With respect to the creature perceived by Frere's friend, however, its height unfortunately frustrates even a Neanderthal identity - unless of course his calculations were in error. Certainly this would be understandable, given the nature of the subject that he had observed at such close proximity whilst alone upon a reputedly haunted mountain! In any event, his sighting is the most cryptozoological of any attributed to An Fear Liath Mor.

Even so, the observer's period of mental disturbance just before its appearance is areoccurring feature of reported encounters upon Ben MacDhui with more phantasmal entities.

One would surely not expect paranormal overtones with material mystery creatures. Yet perplexingly, certain bigfoot sightings, (notably from southern U.S states) have seemingly involved similar psychical components, including an unaccountable fear in observers. Moreover, such fear is also an integral component of folktales worldwide containing ogres and giants.

Could the solution to these enigmas be some retention by modern man of an ancient racial memory recalling an earlier time when our ancestors lived in fear (real or imagined) of Gigantopithecus and other contemporary giant primates? A fear, moreover, that can still be triggered by apparent encounters with, or awareness (pheromonally induced?) of the close proximity of living descendants of these latter collossi?

All most intriguing, but sadly completely speculative in the absence of more substantial evidence on both the cryptozoological and psychological fronts.

Geological holographs.

The Hexham heads are a couple of small stone artifacts, which were found in a Hexham garden in 1972 and have since been associated with a whole series of seemingly supernatural occurrences, including the appearance of a strange wolfman like being, which was accompanied by padding footsteps. In a fascinating and detailed booklet called 'Tales of the Hexham Heads' written by Paul Screeton in 1980, a possible explanation for the wolfman entity postulated by inorganic chemistry specialist Dr. Don Robins is recorded. Upon reading this it became apparent that this theory is in fact equally applicable to the Big Grey Man phenomenon.

Given very briefly, Dr. Robins' theory proposes that various minerals may be capable of storing (encoding) a type of electrical energy, in turn yielding a kinetic image that could actually be released under certain specific conditions. Amongst the supporting evidence that Dr. Robins provides are the visual similarity between the modern-day three-dimensional holograph's and the cup and ring marks from prehistoric times, and the chemical nature of Berthollide compounds - such as iron (II) sulphide and molybendum bicarbide - whose compositions do not comprise simple intramolecular atomic ratios.

Could a geological holograph of this nature be the identity of An Fear Liath Mor? Certainly one could hardly contemplate a more conducive locality for such activity than a rocky mountain peak like Ben MacDhui, and if such minerological manifestations can incorporate both visual and aural components this could surely explain the crunching footsteps and huge figure so commonly reported here. This theory, however, does not answer why Ben MacDhui should yield such a preponderance of petrological phantoms as compared with other British mountains in general. Could such an anomaly be due to certain specific and crucial (but currently undetected) differences in geological composition, historical background, and/or other attributes of Ben MacDhui in comparison to other mountains? It is to be hoped that future researches will provide a key, which in turn may unlock the door to more perceptive understanding of phenomena currently rejected by science as products of trickery and imagination.

In contrast to cryptozoological or geological solutions, other opinions subscribe to a bona fide paranormal identity for the Big Gray Man - a phantom or psychical apparition.

Energy trace theory.

Yet to examine this further, we must first determine the nature of phantoms themselves, and explanations for such are even more numerous than explanations for Am Fear Liath Mor. Some, however, are readily applicable to the Big Grey Man phenomenon.

One of these is called the energy trace theory. techniques such as Kirlian photography appear to reveal that living organisms release intense amounts of energy during periods of extreme emotion - energy, moreover, that can leave behind in a given locality a concentrated invisible trace even after the organism itself has moved away. The energy trace theory proposes that if a locality in which such a trace has been released experiences at some time in the future a similar emotive disturbance, the original trace may develop and actually become visible for a time (comprising the 'phantom' or 'ghost'), comparable in fact to the original invisible image on a photographic film becoming invisible when subsequently developed.

Some mysterious events recorded from Ben MacDhui occurred directly after the climbers concerned had been contemplating the Big Grey Man legends. Could their thoughts, intensified by their solitary, hitherto undisturbed travels upon the mountain, have triggered the appearance of some energy trace image created originally by earlier visitors to Ben MacDhui? The energy trace theory additionally postulates that some people release greater amounts of energy than others, thereby offering an explanation for why many climbers have not experienced anything strange upon Ben MacDhui - namely, that these must have followed people who did not release sufficient energy themselves to create any original invisible trace image.

This thought-provoking theory - which corresponds closely on many points with Don Robins's geologically based proposal - would require, like the latter, an aural component to account for the Big Grey Man's footsteps, and is similarly deserving of serious scientific study.

The more traditional explanation of a phantom - that it comprises a person's spirit - or astral body - has also been applied to the Big Grey Man. Indeed, it has even been identified by some writers! Namely, as the ghost of Cairngorm poet Uilleam Ruighe Naiomhe. Moreover, in a letter to 'The Scotsman' of October 25th 1941, Seton Gordon noted that one apparent sighting of a Ben MacDhui phantom was made by a local inhabitant who, while crossing its plateau to Coire an t-Sneachda one stormy day, realised that a pallid man was following him, wearing only a shirt despite the bleak weather. At the Coire's edge, this man simply disappeared! Such a totally human phantom, however, contrasts starkly with the guise generally attributed to the Big Grey Man - a huge, long limbed, often mist-shrouded entity, sometimes seemingly non-human - hence it is probably unrelated to the latter phenomenon.

Electromagnetic phantom.

Conversely, a hypothesis which may be of notable significance to An Fear Liath Mor is the compelling suggestion that phantoms could be energy forms occupying the immediate boundaries of electromagnetic radiation, i.e the Infra Red (IR) and Ultra Violet (UV) regions. For there is incontravertible evidence that certain ghost like images unseen to photographers have been captured on film sensitive to these wavelengths, as affirmed by researcher of psychical phenomena. Andrew Green relative to IR and George Kanigowski of the Nottingham based Strange Phenomena Research and Notation Group (S.P.R.I.N.G) relative to UV. ('The Unknown' February 1987).

According to Andrew Green, for example, NASA have testified that they have obtained locality photographs that reveal ghostly heat (IR) images of people and vehicles which left the locality many hours before it was photographed. Heat (IR) images may explain many 'photographed phantoms' - especially those in contemporary attire.

Could the phantom of the UV wavelengths be the genuine article - real ghosts? According to frequent documentation in literature appertaining to paranormal phenomena, investigations of notable psychics have often revealed that their eyes' colour sensitive perception is displaced to the right of that of everyone else. That is to say, the eyes of psychics appear less sensitive than those of other people to the red zone of the visible spectrum but overly sensitive to the violet extreme and seemingly even beyond that, into the UV (astral) zone (normally invisible to human eyes). If this is indeed true, and psychical phantoms (whatever they may be) reflect UV light, this would account not only for their visibility to psychics and invisibility to everyone else, but also for many examples of mysterious photographs which clearly depict forms unseen by the photographer but which are evidently something more than mere heat images.

Could the Big Grey Man be an electromagnetic phantom - an entity beyond the vision of most humans (hence the rarity of sightings compared with the frequent reports of footsteps), but whose presence is still sensed?

What might be revealed one day, I wonder, by a climber on Ben MacDhui who points a camera containing UV sensitive film in the direction of crunching footsteps being produced by an apparently invisible agent?

Optical illusions.

Many reports of phantoms have undoubtedly resulted from optical illusions. We can all recall incidents in which, following closer observation, we have perceived an object to be totally different from what it initially appeared to be.

Such frequent occurrences are due to our brains making assumptions concerning the appearance or identity of a poorly or briefly seen object, and then 'filling in' the missing details, so that we think that we have seen the object in greater detail than is really the case. Numerous articles examining such occurrences have been published, but one of the most interesting is that of R.A.Proctor, entitled 'Notes on Ghosts and Goblins' (Cornhill Magazine, April 1873). One of many examples contained in it featured an incident from his own experience. Very distraught after his mother's death, he awoke one night to perceive, standing at the foot of the bed in his moonlit college bedroom, a white image of his mother, perfect in every detail. Moreover, he readily discerned tears sparkling in her eyes. Transfixed, he lay gazing at the image for some time, then leaned forward to observe it more closely. Instantly it transformed itself into his college surplice hanging down from its hook, with the tears becoming the silvered buckles of a rowing belt that he had hung over his surplice! Of great interest, furthermore was the fact that when he deliberately lay back in his original position in an attempt to revisualise the illusion of his mother's image, he found that he could not. His surplice and belt remained as such in his eyes. For now that his brain was aware of their true identity it could not be fooled into 'filling in the details' to create the image of his mother again.

Another form of optical illusion that involves this principle, and which is especially pertinent

Brocken Spectre
From the collection of Dr Shuker

to the Big Grey Man phenomenon, is known as the autokinetic effect. As described by Nigel Watson in 'Fortean Times' (Autumn 1982), this illusion occurs when there is a lack of visual clues for an observer to detect an object's movement (or absence of movement).

The result of this is the observer's perception of apparent movement by the object concerned when in fact the latter is completely immobile. Climber W.H.Murray gave a good example of the autokinetic effect in his book 'Mountaineering in Scotland'. Whilst in the Cairngorm's one day, he saw what he believed to be two climbers traversing from east to west across the snowscape a long distance ahead of him. As he approached these 'climbers', they paused, apparently waiting for him. Ultimately, however, when he was about 100 yards away, he perceived their true identity - they were a pair of black boulders! Clearly as Dr. Helen Ross, comments in 'Behaviour and perception in Strange Environments' (1974), the autokinetic effect can evidently explain certain sightings of phantom-like figures and giant beings apparently observed upon desolate, mist-shrouded mountains.

Hallucination or Brocken Spectre?

As noted earlier, since Professor Collie's Am Fear Liath Mor account, many mountaineers have contemplated the rumours surrounding the Big Grey Man whilst actually climbing Ben MacDhui, and even those who have not done so have been for the most part well aware of the phenomenon. Consequently, given a sudden unexpected sound, or strangely shaped wall of mist, it could be readily envisaged how the startled minds of such climbers could 'fill in the details' to create - especially if further influenced by autokinetic illusions too - a very real (to them) phantom or presence they would sincerely believe to be the Big Grey Man!

Another form of spectre that COULD be involved in the Big Grey Man phenomenon, however, is one whose origins are neither psychical or psychological. Under certain climatic conditions in mountainous regions, a person's shadow is very greatly magnified, and is sometimes thrown onto a cloudbank or patch of mist. This eerie effect is sometimes known as a Brocken Spectre, named after a peak in the German Harz Mountains from which the effect was first recorded. It has frequently been suggested that this could be responsible for various sightings of huge beings upon Ben MacDhui and associated terror. Certainly it can be an unnerving sight to the uninformed observer; so too can be the weird shapes sometimes produced by local condensations of water vapour. Naturally, however, neither of these could explain the aural components of Big Grey Man encounters. Certain precipitation effects could, but conversely these could not explain the sightings.

Benevolent.

Equally problematical are reports of meetings upon Ben MacDhui with benevolent beings - totally contradicting the threatening, sometimes distinctly malign, persona normally associated with Am Fear Liath Mor.

In an *Aberdeen Press and Journal* letter of November 30th 1925, its anonymous writer suggested that the Big Grey Man was a Deva - a nature spirit. Devas have been reported from many parts of Britain - and in recent years - including the Lake District, the Irish peak of Slieve-na-mon. and especially Findhorn in Scotland (see Paul Hawken's detailed book *The Magic of Findhorn*. Being concerned with maintaining the natural beauty of their domains Devas apparantly are sometimes quite hostile to human intruders. Paradoxically, therefore, if this benevolent identity is applied to the Big Grey Man, it could actually explain the latter's belligerence towards climbers on Ben MacDhui. In *Fairies at work and, at play* (1925) Geoffrey Hudson. a noted Theosophist, described a winged Deva he had seen upon a hill in the Lake District as being between 10-12 feet in height and radiating a raw, powerful force - features shared by the Big Grey Man. Is Am Fear Liath Mor some form of elemental?

Enchanted Britain author Marc Alexander has wondered whether the Big Grey Man could be a urisk the offspring of a human-fairy union, supposedly dwelling in solitude within remote mountain regions of Scotland.

Practising Mahayana Buddhist, Sir Hugh Rankin has a very different belief, namely that the Big Grey Man is a Bodhisattva - one of the five 'perfected men' controlling Earth's fate. Moreover. Sir Hugh and his wife informed Affleck Gray that they had actually seen and spoken with him. A similar belief is also held by the Rev. Countess of Mayo, a mystic and leader of Edinburgh's Active Truth Academy, who suggested to Mr. Gray that 'The Golden Big White Men of Ben MacDhui would be a more appropriate title for his book. Unlike the Deva theory, however, this cannot be reconciled with Am Fear Liath Mor's forbidding mien.

In 1953, prominent science-fiction writer Isaac Asimov published a short story entitled 'Everest', in which the yetis turned out to be benevolent Martians observing mankind from the seclusion of the Himalayas. A comparable scenario has been put forward - but this time wholly on a serious level - for the Big Grey Man. Its principal proponent is Rev. Dr. George King, founder and leader of the Aetherius Society, who claims that Ben MacDhui is inhabited internally by highly advanced extra-terrestrial beings (termed the Great White Brotherhood). engaged in producing a complete assessment of Mankind's total spiritual work upon earth. Once again, however, this theory is not compatible with the predominantly malign demeanour associated with the Big Grey Man.

Composite.

Though very diverse, all of the identities discussed within this article share one common feature - none can wholly explain the body of equally diverse events recorded from Ben MacDhui that have been attributed to the Big Grey Man.

In fact, Am Fear Liath Mor is a Big Grey Man no longer. It has become a phantasmogorical phalanx of phantoms, a veritable vortex of variations - each more enigmatic than any penned by Elgar!

Assuming that all of the varied reports are indeed valid, it is clearly inconceivable that a single phenomenon is involved. Only one solution is tenable - that the 'Big Grey Man' supposedly responsible for these is a composite - created by the wholesale aggregation in the past of several completely separate, mysterious phenomena to yield a single artificial, polyphyletic unit labelled with the name of the best known of these component phenomena - namely Am Fear Liath Mor.

Thus, to rediscover the true Big Grey Man, we must weed out those reports that are evidently extraneous. The benevolent beings, the hairy solid biped, John Grant's quadrupedal horror, the sourceless music; all of these, though equally mysterious, are clearly unrelated to that which remains - the well defined, unvarying entity encountered by Professor Collie, Dr. Kellas, and various others. Namely, the real Big Grey Man, Am Fear Liath Mor - the huge, phantom-like being, often unseen but usually sensed, producing long, striding footsteps and inducing terror in those who encounter it.

Even so, two fundamental problems still remain. Why does Ben MacDhui house such a plethora of inexplicable phenomena, and what (if any) is the function or relationship of the Big Grey Man within this perplexing panorama of the paranormal?

In addition to Ben MacDhui, many other localities in Britain and elsewhere appear to possess mystical attributes, and to stimulate psychical sensations that seem linked not to the people who experience them, but to the localities themselves, as noted by Marc Alexander. Some are explained away as the result of a Ley Line conflux, i.e arising through the emanation of earth energies. In various other cases, however, such locations are popularly held by researchers into the paranormal to be window areas - portals through which visitors from other dimensions can enter our world's time and space. Could Ben MacDhui be one such window area'?

Such a highly significant locality - an interdimensional interface - would surely be monitored and protected by some sort of guardian, whose function would be to warn off any potential intruder (deliberate or accidental), ensuring by every means available to it that everyone kept well away from the portal's vicinity (in this case apparently sited on Ben MacDhui's peak). It can be seen that the above description corresponds perfectly with the reported activity and behaviour of the Big Grey Man!

Is the above scenario possible? Are any of the identities discussed here relative to Am Fear Liath Mor near to the mark? Who can say? It would appear, however, that Ben MacDhui does possess a number of unsolved mysteries - including an entity that apparently seeks most emphatically to keep them that way!

* * * * *

The Melanotic Leopards of Eastern Cape, South Africa.

by Chris M Moiser.

In August 1995 I visited Cape Town with David Barnaby. The primary purpose of our visit was to look at the progress of the Quagga Project. This is the attempt to recreate the partially striped zebra that became extinct in the latter part of the nineteenth century. The project centred on the South African museum in Cape Town which, quite appropriately is the only museum in Africa with a mounted Quagga. Indeed, in part it was the remounting of this specimen by Reinhold Rau, the head taxidermist that led the series of events which culminated in the tart of the Quagga Project.

On our first day in the museum, whilst finding the Quagga, we viewed the other mammal displays generally, and it was at this time that I found the brown leopard! It was a reasonable leopard size, with a general dark overall tawny-brown colour which gave way to a paler cream on the neck, face and back leg. This creamy colour was covered in spots in the dark tawny brown. The spots were up to 10mm across and they were spots not rosettes. Where the skin was brown it was pure brown with no rosettes showing through as they do typically in the black leopard.

The animal was in a case with a Caracal to its left, a Lion to its right and a King Cheetah above the lion. This cat case was directly opposite the mounted Quagga, and in a nearby gallery was a cast of the original Coelacanth that was caught in 1938. My interest in those strange and enigmatic species which have always fascinated cryptozoologists increased many fold in a matter of seconds.

Having spent most of my spare time of the previous year investigating and recording Alien Big Cat sightings in the south-west of England, this brown leopard was almost a dream come true. It fitted the description of so many cat sightings that I had heard. A brown leopard length cat, with a leopard length tail. Sadly, as subsequent investigations were to reveal, there is no real likelyhood of leopards of this variety ever having been taken into captivity alive, let alone being imported into the United Kingdom and escaping.

It was a beautiful animal, and not just because of its novelty value. Time on the trip was at a premium with our many other commitments so most of the research was going to have to be done from the United Kingdom by post. Much to the museum's embarrassment they couldn't find their records on the animal. We did, one afternoon, even open the case to see if a record card had been secreted near the animal, a practise which sometimes happens in museums. We found nothing, but liberated a large volume of preservative fumes into the mammal gallery.

South Africa showing the Eastern Cape

At some time in the past I had read 'Mystery Cats of the World' (Shuker, K.P.N. 1989), and on my return to the United Kingdom I contacted Karl Shuker who was able to supply me with a couple of references. At the same time I also wrote to the Albany Museum in Grahamstown, the place where Karl assured me these animals had first been recorded.

The information then began to accumulate quickly, in a greater volume than I had ever hoped for.

The first report in the scientific literature seems to appear in 1883 when R.Trimen published a short report in the Proceedings of the Zoological Society. He reports seeing a skin in the Albany museum at Grahamstown. This specimen was apparently presented by a Mr Buckley in August 1870, having been shot at 'Bucklands, near Koonap'. The museum catalogue is said to have described this animal as a cross between an ordinary and a black leopard. Trimen expresses his doubts to this, though, on the quite reasonable grounds that he believed the black leopard to be unknown in South Africa. From the description he gives it seems that this specimen had more spots than the Cape Town specimen.

Dr. A Gunther then comes up with two reports (Gunther A 1885 and Gunther A 1886). In the first the flat skin of a 'rare leopard' has come into his possession. The animal had been killed forty miles from Grahamstown, in a hilly district covered with dense scrub. Gunther tentatively suggested that it might be the result of a cross between a Leopard and a "Lioness which had strayed that far South". (The Lion had been absent from that area "for a considerable period"). He then goes on to compare the animal with the Asiatic black leopard, and suggests that this animal is an example of incipient melanism. It is here that the word melanotic is first used.

The second report (Gunther, A. 1886) is largely based on a letter that Gunther received from the Reverend Nendrick Abraham, the president of Grahamstown's Natural History Society. Abraham had his own skin and enclosed a photograph of it, but with a warning that the photograph gave no indication of the denseness of the black, and the beautiful gloss. He described his specimen as having been collected about twenty miles from Grahamstown at a place called Collingham. In fact this distance was over estimated - Collingham being about 15km east of Grahamstown. Abraham reported four preserved specimens and had reports of two other similar animals which were living in the area at the time. Abrahams' specimen was presented to the British Museum and is listed as having been accessed on the 15th July 1886.

After these three initial reports information really only exists either as small local records, or as secondary sources, with little in the way of scientific literature until 1987, when Dr. Jack Skead, a former director of the Kaffrarian Museum in King William's Town published a major review work on the "Historical mammal incidence in the Cape Province". In the appendix to volume 2 he devotes eighteen pages to these animals describing them as "the so called black or melanistic animals of the Eastern Cape". In this paper he listed nine documented specimens, either as mounted animals or flat skins. Not all of them still exist.

Photograph by David Barnaby

The case of stuffed felids in the South African Museum, Cape Town
Picture © Chris Moiser

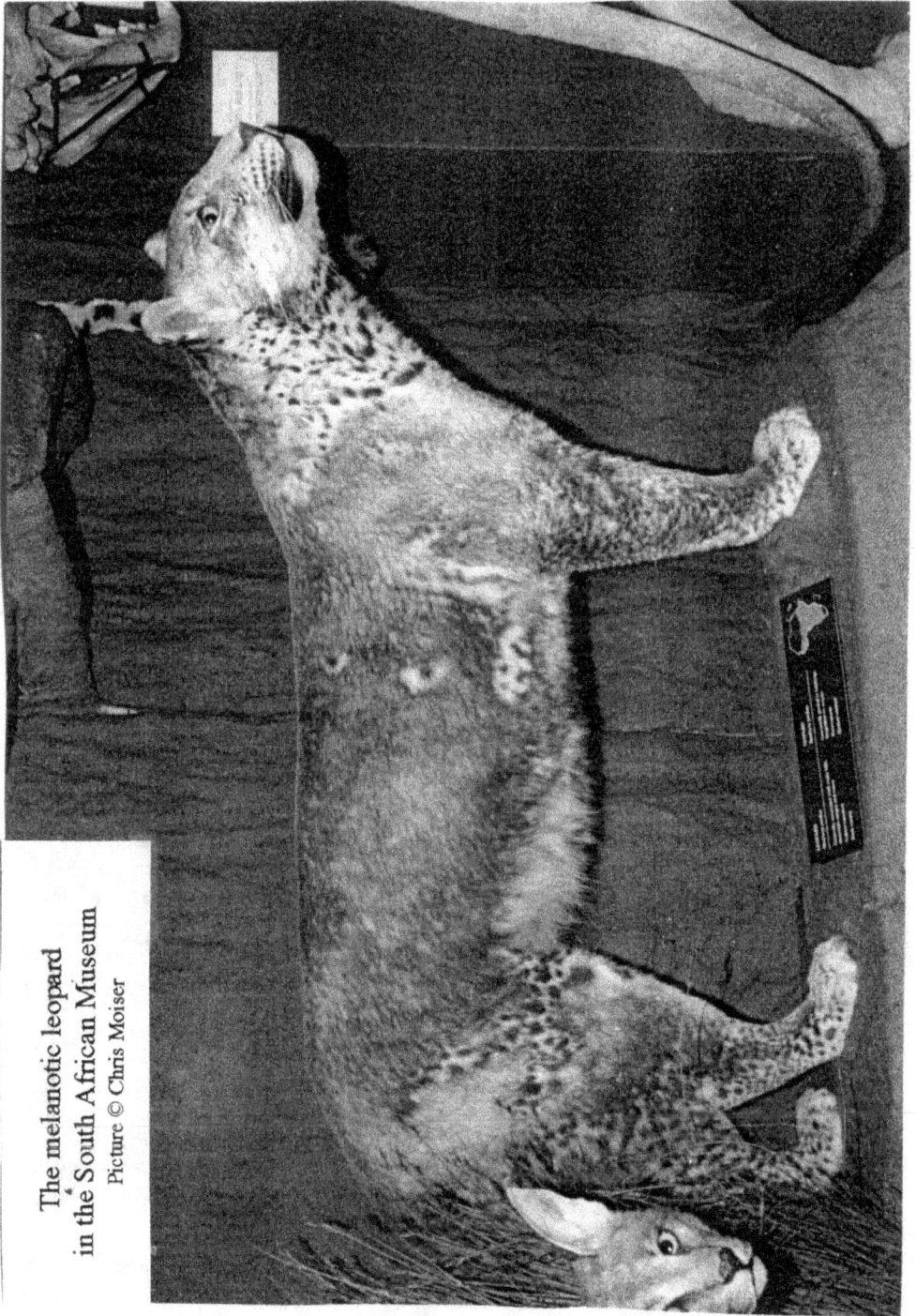

The melanotic leopard
in the South African Museum
Picture © Chris Moiser

The Albany Museum specimen that Trimen (1883) referred to was joined there by a second specimen in about 1900: it is recorded in the museum annual report for that year. Sadly, the museum had a bad fire in 1941, which destroyed one of the specimens and damaged the other. It is thought that it was the second specimen that survived. Interestingly the surviving specimen was remounted in the 1960's and was found to have a lion's skull!

The first of the specimens that Gunther refers to is in the British Museum (Natural History), and is listed as the type specimen for *Felis leopardus var. melanotica*. This is the animal collected forty miles north-east of Grahamstown. The second skin present in the British Museum is the one which was presented by the reverend Abraham.

Abraham also mentioned a skin sold at a church bazaar in Grahamstown. This was prior to 1886, and it is possible, as it seemed to disappear then, that this skin is one of the ones that was eventually acquired by one of the South African museums.

Animal number five in Skeal's list is described in the *'Kaffrarian Watchman'* for 16th October 1889 as "a black tiger" which was shot "a few days ago" in the Idutywa district of southern Transkei, after killing a native. The disposal of this skin was never traced and the only colour description is.. *"this tiger was almost black with a few yellow marks"*.

(The use of the word tiger for any type of large cat was quite common in many parts of the world where tigers do not occur).

The records of the next animal are from a farmer's diary dated 7th September 1891. The diary referred to the farmer, a Mr. Amm, having found *"a large male tiger dead; the black sort"*. This followed the poisoning of the remains of a partly eaten calf that had been found the previous day. The animal was skinned and the skin was used as a rug in the farmhouse until it was destroyed by dogs.. The farm was 18km west of Grahamstown in the Assegai river valley.

The specimen in the South African Museum in Cape Town was purchased from a professional taxidermist based in Grahamstown in November 1898. It was apparently collected 25km south of Grahamstown. The taxidermist was also responsible for mounting the two Albany Museum specimens.

The final specimen existed in the Port Elizabeth Museum but is now lost. It was thought to have been shot in the Humansdorp area at some time prior to 1906. The skin suffered vermin damage but the head remained for some time after the rest of the skin was discarded.

A further black leopard skin exists in the Kaffrarian Museum in King William's Town. This one though is from outside the area and upon investigation it was found to have been imported by Rowland Ward, the taxidermist in London, having come originally from Kenya. Lloyd Wingate, the curator of mammals at the Kaffrarian Museum, describes the specimen as a dark, blackish brown colour, with the spots showing through the blackish brown.

Other records are rare, but they do exist. There is no longer any corroborating evidence, however, for any of them. When the existing specimens and the three known to have been destroyed are considered, there were at least eight of these animals between 1870 and 1891. The actual number of animals which existed would almost certainly have been higher, as the figure of eight assumes that:

i. The specimen *'sold at the church bazaar'* reappeared later as one of the museum skins.

ii. The two living specimens "in the veld near Grahamstown". reported by Abraham in 1886 were later killed and became other described specimens.

iii. No specimens lived and died without being acquired by local collectors or museums.

When the periods during which the animals existed are examined closely, the first specimen, an adult, was present in the Albany Museum in 1870, and another animal was definitely poisoned in late 1891 - a period of 21 years apart.Although a longevity of 23 years has been reported for the leopard (Green, 1991), it is likely that this would have been in a captive animal, and it is likely that wild animals would have a relatively shorter life. This would suggest that there must have been two generations of these animals. If the specimen that was at the Port Elizabeth museum was freshly collected before being exhibited in 1906 this would put the last surviving animal into 1905 at least, giving a time span of 35 years or more for this variety of leopards. The length of time could be suggestive of three generations.

That these animals were nothing more than an abnormal colour variation of the leopard (Panthera pardus) is now beyond doubt, but the colour variation is obviously one that has not been seen since. From the descriptions given in the various papers there is some variation between the individual specimens, but none appear - other than the dark colouration, to be particularly similar to the more common Asiatic black leopard.

With the increasing human population and expanse of farming into Eastern Cape, it seems very unlikely that any Melanotic Leopards still exist, not probable that they have existed since the turn of the century. Certainly, I am not aware of any sightings. The most likely explanation for their existence is that a mutation occurred in a local leopard producing the partial melanism that has been reported. These animals were possibly not biologically disadvantaged in any way, indeed they possibly had evolutionary advantages as several appeared in the local population quite quickly. Their abnormal appearance though, did attract the attention of many of the local hunters, who would normally have treated leopards as vermin and probably treated these animals as vermin with a potentially more valuable pelt than usual!

Acknowledgements

I would like to thank David Barnaby for inviting me to go to South Africa with him, and Reinhold Rau for his hospitality and friendship there. Andre Busch struggled to open the cabinet with me. A very big thank you is due to Dr. Billy De Klerk at the Albany Museum, Grahamstown, who contacted a number of people in the area on my behalf. Lloyd Wingate at the Kaffrarian Museum was also a great help.

References.

Barnaby D, (1996). *Quaggas.and Other Zebras*. Basset Pubs. Plymouth.

Green, R. (1991). *Wild Cat species of the World*. Basset Pubs. Plymouth.

Gunther, A. (1885) *Note on a supposed Melanotic variety of the Leopard from South Africa*. Proc. Zool. Soc. Lond. pp 243-245

Gunther. A. (1886) *Second Note on the Melanotic Variety of the South African Leopard* Proc. Zool. Soc. Lond. pp 203-205

Shuker, K.P.N. (1989) *Mystery Cats of the World - from Blue Tigers to Exmoor Beasts*. Robert Hale, London.

Skead, C.J. (1987) *Historical Mammal Incidence in the Cape* Province. Chief Directorate Nature and Environmental Conservation, The Cape of Good Hope, Cape Town.

Trimen, R. (1883) *On a remarkable variety of the Leopard (Felis pardus), obtained in the East of Cape Colony*. Proc. Zool. Soc. Lond. p. 535.

* * * * *

The Migo is (probably) a Crocodile.

by Darren Naish.

Regular readers of *Animals & Men* will by now be quite familiar with recently-made video footage purporting to show the migo or migua, a lake dwelling cryptid from Lake Dakataua, New Britain. Filmed on a Japanese expedition of January/February 1994, the material, the material was screened as part of a Japanese TV documentary devoted entirely to the existence of the migo [1]. As yet, to my knowledge no part of this footage has been screened on British television: furthermore only a handful of copies (if not less) exist in this country. Among these is that sent to Jonathan Downes by Tokuharu Takabayashi [2]. In 1995, Jonathan Downes invited me to a private viewing of the documentary, and herein I present my conclusions.

EDITOR'S NOTE: As we are just about to go to press with this volume, Rapido TV have just completed their series Fortean TV for Channel Four in the UK. At various times during the pre-production process the possibility of including the TBS Migo footage as part of the series was discussed. Whether or not it will be included remains to be seen.

Previous interpretation of the migo.

The migo has a lengthy and well reviewed cryptozoological history, beginning with W.T.Neill's 1956 report [3], of crocodiles seen in the lakes of New Britain (although Neill did not mention Lake Dakataua). As Neill was not sure as to which species of crocodile he had seen, he suggested that they might be 'an undescribed relative' of the New-Guinea crocodile *(Crocodylus novoguinae)*. Presumably as a result of this suggestion, Heuvelmans included the migo in his 1986 list of cryptids, classifying it as 'an unknown species of crocodile (or is it, as has been suspected, a surviving mosasaur?)' [4].

Mosasaurs are Upper Cretaceous marine lizards, closely related to monitors and snakes, but with flippers and a laterally flattened, sculling tail. The suggestion that the migo is a mosasaur appears to owe its origin to Shohei Shirai, the then chief of the Pacific Ocean Resources Institute, who commented on migo sightings in 1972. [5]

Admittedly the sightings Shirai had in mind were rather fantastic and do not recall the animal captured on film, A mosasaur identity also seems to have been favoured by the Japanese makers of the migo documentary. They included a poor animation sequence of a silhouetted mosasaur in the film, as well as, even worse, sequences of a rather inflexible robot mosasaur. Downes has compared this object to a 'clockwork newt' [6] though I must say that it did not look that bad (it was, however, highly inaccurate).

Equally as improbable as the idea that the migo is a mosasaur is the tentative suggestion of Young and Rosemblatt [7] that it is a 'descendant of the prehistoric *Phobusuchus'*. *Phobosuchus* is a junior synonym of the Upper Cretaceous North American crocodile *Deinosuchus* - an eleven metre, six ton [8] freshwater predator that was .almost certainly adapted for feeding on dinosaurs. The suggestion that *Deinosuchus* may be alive and well and living on an island off New Guinea does not warrant any consideration.

The most popular of non-Japanese theories is that the migo is mammalian, and a surviving 'archaeocete' whale. Roy Mackal acted as scientific consultant to the Japanese expedition - he is even featured in the documentary - and Shuker has stated that, in correspondence, Mackal has ruled out Crocodilian and other candidates, and believes the migo to be an 'evolved archaeocete'. [9] Shuker has concurred, (though Mackal has now changed his mind, see below), and in two articles has suggested that the migo is a descendant of Eocene protocetid whales such as the Pakistani form *Ambulocetus* (news of the migo footage came too late to be discussed at length in Shuker's book *In Search of Prehistoric Survivors* [1] [10]. As discussed below, reasons for considering the filmed 'migo' as mammalian rather than reptilian are, in my opinion, based on misinterpretation. However, Shuker refers to 'short black hair' reported in migo sightings from the 1970s, and has used this feature to endorse his belief that it is mammalian.

This may be reasonable, but is not relevant to the animal filmed by the expedition team. Adding to this confusion is Paul LeBlond's assertion that the Japanese expedition team have suggested an archaeocete whale as migo identity also [11] Unfortunately, LeBlond seems to have identified the mosasaur in the documentary as an archaeocete.

Though speculations reconciling the migo with an aquatic mammal may be valid in that they are drawn upon eyewitness evidence, they do not relate to the Lake Dakataua animal captured on film. This does not correspond to any mammal, living or extinct, and there is no indication whatsoever that it is a whale of any kind. To deduce the identity of the animal in the film though apparently not the migo of cryptozoological lore - we must analyse the film in detail.

EDITOR'S NOTE: As the first person in the UK to see the film much of the initial publicity given in this country was done by me. I sent Karl Shuker a copy. Unfortunately at the time my video duplicating equipment was primitive in the extreme, and he was unlucky to get a copy somewhat akin in quality to one of the 'bootleg' copies of Disney movies which one can purchase at car boot sales. It appears that my equipment even managed to miss out bits of the documentary, and it has to be stressed that in my opinion, (and indeed that of both Karl and Darren), many of the differences in interpretation between the two of them come from the fact that Darren had access to a far better quality copy of the film!

Migo Morphology.

No-one has yet satisfactorily conveyed the appearance of the migo, as seen in the film. Published diagrams [2] [12] are highly schematic. This is mostly because, though highly animate, the animal is amorphous, some difference from the camera, slightly out of focus, and, most importantly 'highly pixelised' due to digital enhancement. The effect of the latter is that all movements are 'jerky' as material 'moves' from one pixel to the next across the screen, and that areas of similar shade, (including, for much of the footage, the water-object interface) are amalgamated. Smooth surfaces are obliterated and replaced by 'staircase' outlines. Nevertheless, the animal is seen for long enough and with sufficient resolution for a detailed analysis.

Unlike so many cryptozoological cinematic debuts, the migo footage is far from a fleeting glimpse. The animal moves sedately from right to left across the screen almost exactly in parallel with the bank visible on the other side of the lake. It is dark in colour, and very low lying so that a row of dorsal extremities are all that is visible for most of the time. The object does not form a single unbroken area above the water surface for most of the film, as there is frequently water in the gap between the head and the low hump formed by the back (figure 1). I am presently of the opinion that the 11m estimate given by Mackal [1] [15], is far too high - judging from birds that appear both in front and behind the animal during the sequence, it appears that it is not more than 6m and probably less. If, however, Mackal's observations are the result of sightings in the field, rather than later viewing of the video footage, they are harder to dismiss.

As the animal moves uniformly as a continuous mass, corresponds with recogniseable body plans (head-body-tail), and occasionally reveals parts of body between the exposed extremities, the possibility that it represents more than one animal can be rejected.

The head has a distinctive profile and I was able to discern a distinct shape by repeated viewing. It is composed of two sections, a low anterior area, and a raised posterior area which is flat-topped (but recall notes above on effects of digitalisation). This hind area creates the impression of a shallow upturned box set on top of the lower portion of the head. Occasionally a raised bump is discernable at the tip of what must be assumed to be the snout. There is a single black blob at the back of the raised portion. (Figure 1).

Immediately behind the head is a lower area in which, oddly, two vertical spines are seen to project. These are dark, and there are always two (Figure 1). We must assume that they are set upon the animal's neck, and, whilst this part of the body is visible at times during the footage, often the two spines are the only part seen to project above the water surface. At these times there is a water-filled gap between the head and the start of the back. I believe that the spines have been distorted by digitalisation.

Figure One.

The 'migo' as filmed by the Japanese film crew.
An earlier conception of this diagram [14] incorrectly
shortened the 'tail gap'.

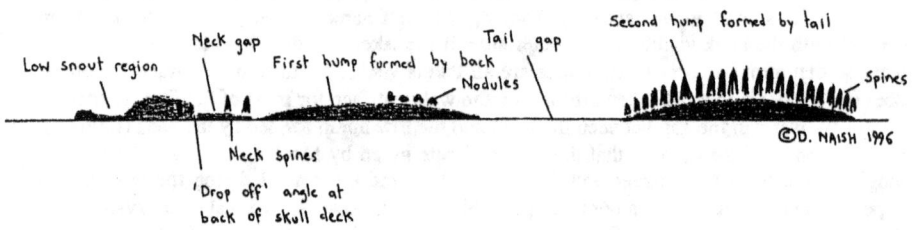

Low snout region
Neck gap
First hump formed by back
Nodules
Tail gap
Second hump formed by tail
Spines
Neck spines
'Drop off' angle at back of skull deck
©D. NAISH 1996

The 'back' of the animal begins after the neck gap. At times, it forms a very low hump out of the water, and has at least four small, dark nodules on its uppermost surface. It does not appear to move vertically in the water. It is approximately three times the length of the head. Toward the tail of the animal, it descends at an angle into the water, and there is then another water-filled gap before we reach the tail (in a version of figure one published elsewhere [14] the gap between body and tail was shown as shorter than it really is).

The tail is the most problematic part of the animal revealed by the footage. It forms another shallow hump, slightly higher out of the water than the back, and has triangular vertical spines along its entire length. It is approximately the same length as the hump formed by the back. The cause of much confusion, the result being the interpretation of the animal as mammalian, is that the tail appears to move vertically in the water. Often it is entirely absent, and then rises above (what appears to be) the water surface.

All of these features are annonated in figure one.

Migo as a cetacean.

My initial interest in the migo and its background, came as a result of claims discussed above, that the animal filmed on location was a whale, and furthermore one of the hypothetical surviving archaeocete whales. As can be determined from the description above, no diagnostic features are evident that would allow this animal to be seen as belonging to any one of the known groups of swimming animals. The allocation of the creature in Mammalia is due to interpretation of what appears to be vertical undulation: an interpretation I contest, and believe to be an artifact due to the medium (see below). Ignoring this for the moment, and suspending any scepticism, let us imagine that the animal described above is a protocetid whale. If I were to interpret the animal filmed by the expedition team as a whale, the beast seen in figure two would be the result. This is clearly a bizarre and an entirely new sort of animal. As explained below, such an interpretation is not the most parsimonious of options, and certainly the least likely.

Unfortunately the cetacean 'status' of the migo has been erroneously supported due to confusion with other animals that can also be seen in the footage. In his reviews of it Dr Karl Shuker [1] [9] has referred to a few seconds of footage which appear to show "two slender projections resembling dorsal fins or spines" and "the vertical emergence of what may have been a tail, with two horizontal whale like flukes". He used these glimpses to endorse his notion of the migo as protocetid whale, and thus they were highly significant - if they really were body parts of the migo they might prove that it was a protocetid whale after all. It was therefore essential that I see them for myself on tape.

With Karl Shuker's assistance, Downes and I were able to locate these few seconds. To our surprise, it seemed likely that these had not actually been filmed in Lake Dakataua at all, but apparently from a boat approaching the island at sea. This appears to correspond with Paul

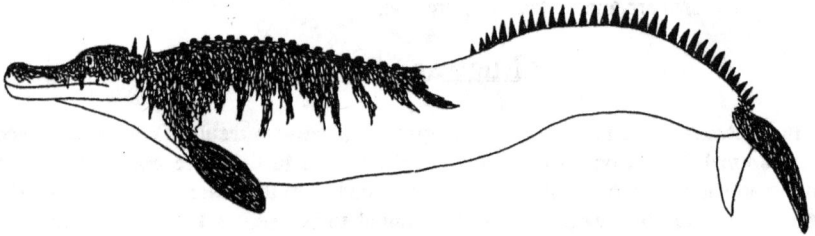

Figure Two.

The 'migo' interpreted as a protocetid whale. An
impossibly ridiculous animal!

LeBlond's commentary on the documentary [11], and is an opinion recently endorsed by Ben Roesch after his viewing of the footage. [15]. As the editor has pointed out, however, the copy of the film originally viewed by Dr. Shuker was so flawed that such misinterpretations of continuity are perfectly understandable!

It appeared therefore that the footage featuring whale like fins and tail-flukes was located within this part of the footage, and was therefore unlikely to be anything to do with the migo itself. It appeared to be film of dolphins. In a sequence lasting only two or three seconds, three silhouetted dolphins (either *Delphinus* or *Stenella* species) were seen to rise and then dive: and as each dolphin dives, it presents a smooth rolling shape with a dorsal fin. As the three are silhoutted and close together, they appear as a single large mass in which, even on a relatively clear copy of the documentary, it is difficult to see where one animal ends and another begins. The first dolphin, somewhat ahead of the other two, dives first, but does not rise high enough out of the water for its tail to emerge above the surface. The second and third surface and dive almost simultaneously. A beak and foreparts of a dolphin are seen for a fraction of a second. With the second dolphin in front of the third (and they are overlapping) there is the brief false image of a single rolling shape with two dorsal fins. Finally the tails of both appear briefly above the surface with the tail flukes of the third being discernable. This sequence is illustrated in figure three.

By proving, therefore, that the footage in which cetacean flukes and fins appear is entirely disassociated from the footage of an animal swimming in the lake, the better part of the case that the latter might be cetacean is removed. We are left with the swimming footage described above. And this is of a crocodile.

Migo as an Indopacific Crocodile.

Having watched and scrutinised the migo footage for myself, and having considerable experience with video footage of swimming crocodiles and whales, I now believe that the habits and morphology of the animal filmed in Lake Dakataua to be almost certainly that of an Indopacific Crocodile.

There is nothing in the footage, bar the apparent vertical movement of the tail, that corresponds with a mammalian identity. This vertical motion is important, if it is real, and if it is considerable because it would expunge all crocodiles from the list of contenders. My initial suspicion, was that the tail was sculling from side to side, as does a crocodile's, but with each sideways sweep was moving slightly vertically, enough for it to disappear beneath the water.

Having spent many hours poring over footage of swimming crocodiles and alligators, however, I cannot find this to be the case, nor are there any indications from the crocodilian caudal skeleton that the bones move vertically as the entire structure sweeps laterally. It would seem instead that the digitalisation is responsible in that lines of pixels make the object-water interface indistinguishable at times, and any slight vertical movement results in material

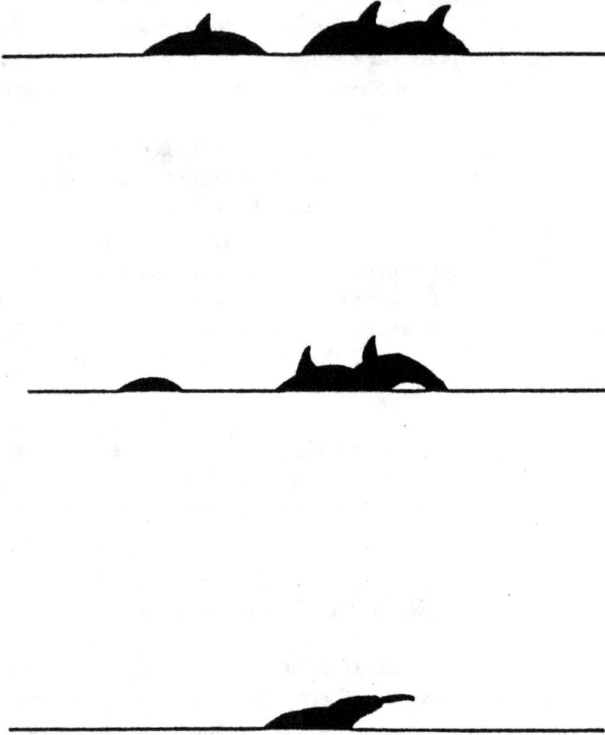

Figure Three.

Highly schematic representation of the 2-3
second dolphin sequence that appears in the same documentary
as the 'migo' footage.

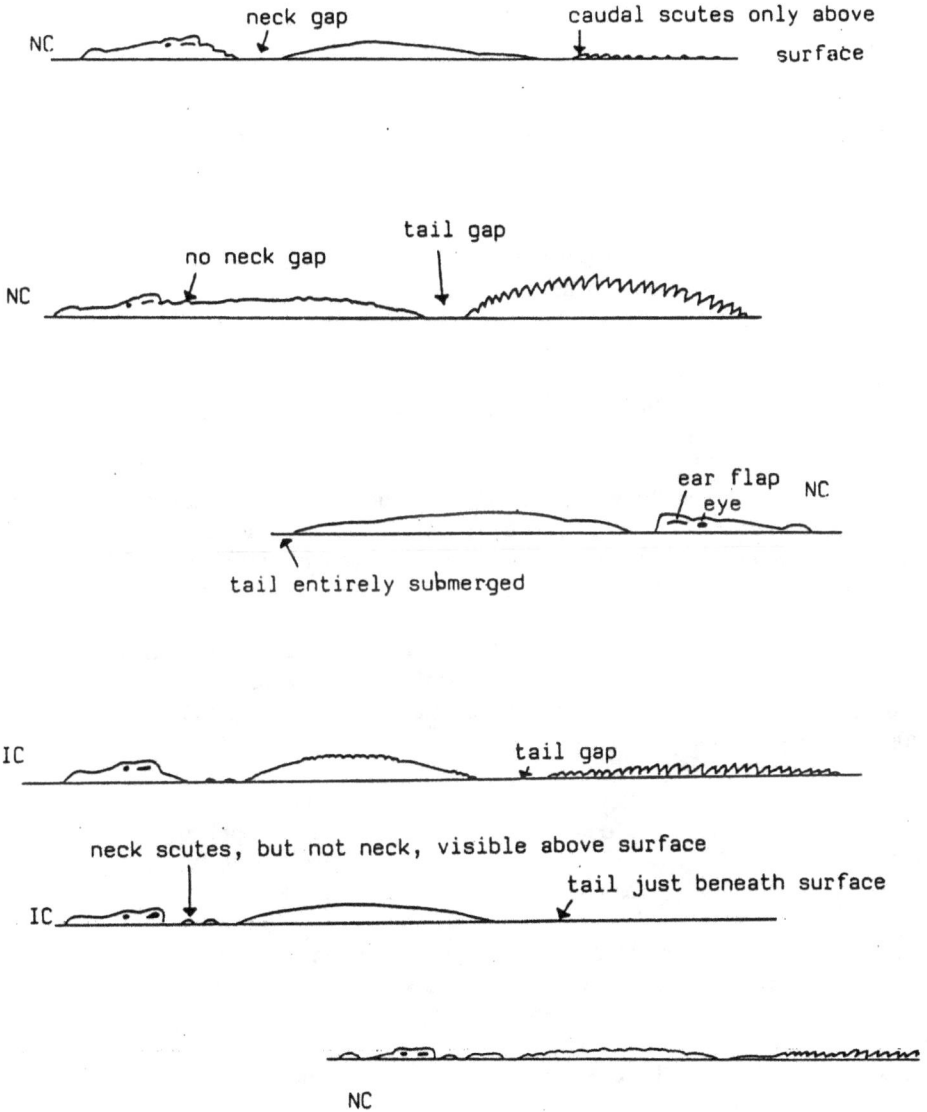

Figure Four.

Schematic descriptions of swimming crocodiles
as seen in photos and on video footage. Those labelled NC
are Nile crocodiles and those labelled IC are Indopacific
crocodiles.

'jumping' vertically from one line of pixels to the next - in this case up and down on the screen. Whether or not this is the case, however, crocodiles do vary their vertical position in the water as they swim along, and often scull with the tail slightly below the surface of the water, as well as above it.

I have seen swimming crocodiles whose tails would match all positions displayed by the filmed individual in Lake Dakataua (figure four).

However, for me, this is not the most important part of the case. In feature after feature, the migo in the documentary footage corresponds exactly with an Indopacific crocodile. As described above, the long head, with:

(1) A raised terminal nostril bump.
(2) A moderately long, low snout.
(3) A hind, box-shaped portion higher than the snout, with a fairly steep 'drop-off' angle at the back (called the 'skull deck').

These all correspond exactly with a *Crocodylus* crocodile. The dark area towards the posterior part of the head appears too far back to be an eye - it may be a blurring of the ear flap behind the eye. This is black and can often be very conspicuous. Some individuals of the Indopacific crocodile have dark spots or stripes, and one of these instead may explain the feature.

Particularly telling, if not diagnostic, are the two vertical 'spines' on the neck alluded to in the description above. It cannot be coincidence that these objects appear in exactly the same place in the Lake Dakataua animal as do conspicuously raised dermal ossicles in the Indopacific crocodile (Figure five).

The bumps and spines seen along the body are also crocodilian in appearance - especially the row of them along the tail. Dermal spines such as these are a reptilian feature - in modern forms, they occur in iguanas and other lizards, in tuataras and throughout crocodiles. In contrast it is impossible to reconcile them with a mammalian identity.

Incidentally, the existence of a row of dermal spines in the migo may have been what originally prompted comparisons with mosasaurs. It is all the more ironic, therefore, that mosasaurs did not posess a row of dermal spines at all!

Traditionally they are depicted with a wavy crest growing along the entire length of the back and tail (Figure 6a).

However, the only piece of evidence for this turns out to have been a misplaced piece of tracheal cartilage [16]: It seems instead that mosasaurs were sleek and crestless. (Figure 6b). Crests along the backbone or not, it cannot be emphasised enough that mosasaurs are irrelevant to the animal filmed at Lake Dakataua.

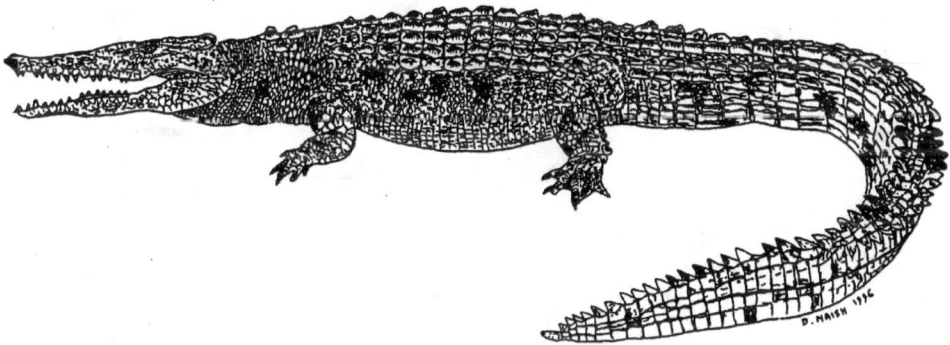

Figure Five.

An Indopacific crocodile *(C.porosus)*. Note the unique
pattern of osteoderms on the neck. Typically between
4-6m long when adult.

©D. NAISH '96

Figure Six.

The changing conception of the Mosasaur.

6a. A 'traditional' restoration of the European mosasaur
(Mosasaurus hoffami) based on a famous painting by Zdenek
Burian. Note the dorsal fin and the crocodile-like scutes
and nodules.

6b. An up-to-date conception of *Clidastes propython* (a
smaller type of mosasaur):This is a sleek, streamlined beast
with a seal-like profile.

Expanses of water appearing both between the head and between back and tail of the Lake Dakataua animal were noted in the description. They are also characteristically crocodilian. They can be seen in virtually every surface-swimming crocodile. Swimming crocodiles often have their heads lifted relative to the body, and with sloping forequarters they create a submerged area just behind the head (fig 4). The base of the tail can also be shallowly submerged while the rest of it, namely that bearing the serrated outline of dermal scutes, is visible above the water (fig 4). I have seen video footage where the crocodile's above-water profile corresponds exactly with that of the migo.

In all of its rather precise details, therefore, the animal filmed in Lake Dakataua corresponds with what we see in swimming Indopacific crocodiles. Downes has suggested to me that the migo's large, barrel shaped body is at odds with this identity [17]. However, large Indopacific crocodiles can be wide, bulky animals with a body that could be described as 'barrel shaped'. I am almost totally certain that the animal filmed at Lake Dakataua is, after all, an Indopacific crocodile. This opinion has now been independently reached by Mackal. While this article was in preparation, I received an email from fellow researcher John Moore informing me of Mackal's present thoughts on the migo. Besides thinking that the migo is, after all, an Indopacific crocodile, Mackal also explains the presence of a second migo film, this time of a smaller animal swimming closer to the camera. According to Mackal this is a crocodile as well [18].

Crocodiles in and around the Bismarck Archipelago

We have already seen now Neill saw crocodiles in lakes of New Britain, and suggested that they might be of an unknown species. Though this is within the realms of possibility, it is not the most likely option given that New Britain is already within the extensive range of the far-reaching Indopacific or Saltwater crocodile (C. porosus).

As is well known, Indopacific crocodiles are able to exploit the marine realm by virtue of their effective salt glands, and are highly mobile animals. Found throughout tropical regions of Asia and the Pacific (figure seven), their range extends from Sri Lanka and the southwestern coast of India to northern Australia and New Guinea, and across the Indonesian and Philippine islands. To the east they have been known to reach the Solomon Islands, Vanatu (= New Hebrides), and Fiji in the south [19], and Pohnpe (= Ponape) in the eastern Caroline Islands in the north. [20] The western record seems to be of a solitary individual in the Cocos Islands, but the most extralimital record is surely of an individual observed 48km from New Zealand's North Cape. [19] Further extra-limital records are those from southern China [19] and Hong Kong, as hinted at by local place names and mythology [17].

EDITOR'S NOTE: My colleague Richard Muirhead has been researching reports that a definite record of this species from the territorial waters of Hong Kong was made in 1912. There are several references to this record throughout the literature but the original record remains elusive!

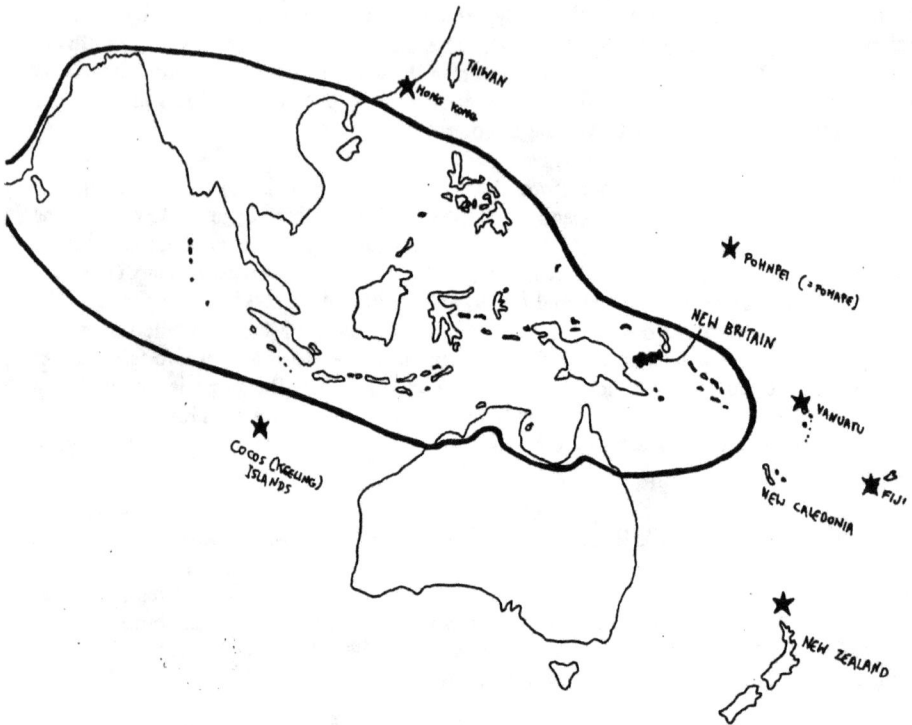

Figure Seven.

The extensive range of the Indopacific
crocodile(the total area within the bold line). Extra
territorial records are marked with a star, and New Britain,
the home of the migo has been blacked in.

New Britain is therefore well within the range of the Indopacific crocodile. The alternative name 'Saltwater crocodile' is a misnomer as this species occurs widely in freshwater lakes and rivers. Though largely restricted to downstream of major waterfalls or barriers [21], it has been recorded considerable distance inland - as much as 1130km inland in New Guinea [22] [23]. Though I have yet to discover a record stating the presence of Indopacific crocodiles in Lake Dakataua, we can infer that, in all likelyhood they occur here. Allen even refers to 'known populations' of *C.porosus* in the Bismarck Archipelago [20].

Conversely, the mostly freshwater *C.novaeguineae* has a rather more restricted range and has not, apparently, been recorded from New Britain. It can be excluded from candidature by virtue of its size: It is generally around three metres long, but can reach four metres. Its scalation pattern is also different from that shared by both the Indopacific crocodile and the Lake Dakataua animal (if pixelised it would not have two vertical 'spines' behind the head).

Of the several other crocodile species known to occur in the Indopacific region, none are found west of New Guinea. *Crocodylus porosus* is thus the only known species that answers to the Lake Dakataua animal.

Conclusions.

1. The appearance of the Lake Dakataua animal corresponds with that of a large crocodile, and almost certainly with that of *Crocodylus porosus*.

2. None of the characteristics of the Lake Dakataua animal suggest a mammalian identity. In view of its resemblance to Indopacific crocodiles, it is simply illogical to suggest an alternative identity.

3. Footage of delphinid cetaceans is included in the Japanese documentary about the migo. This should not be confused with footage filmed at Lake Dakataua - footage which shows a large crocodile swimming in the lake.

4. New Britain is within the range of the Indopacific crocodile and of no other large crocodile species.

Acknowledgements.

Special thanks to Jonathan Downes for private viewing of the migo footage, to Karl Shuker of assistance and above all to Tokuharu Takabayashi for kindly sending the tape to the CFZ in the first place. Thanks to Paul Willis for information on swimming crocodiles, to Tim Isles for providing video footage, to Greg Paul for providing help with a reference, and to all members of dino-1 who helped me with queries on mosasaur skin, in particular Gary Kerr, Tracy Ford and Tony Thulborn. Thanks to Colin McHenry for discussion on the finer points of crocodile ethology, to Peter Edwards for more swimming crocodile footage, to Aaron Venables for loan of skeletal material, to Ben Roesch for feedback and comments, and to John Moore for comments and information on Mackal's views. Ain't zoology cool?

References and Notes.

1. SHUKER, K.P.N. 1995. *The Migo Movie: A further muddying of murky waters.* A&M 5: 22-25.

2. DOWNES, J.1995 *Crocodile Tears II.* A&M 5:22

3. NEILL, W.T. 1956. *The possibility of an undescribed crocodile on New Britain.* Herpetologica 12: 174-6.

4. HEUVELMANS, B. 1986. *Annonated checklist of apparently unknown animals with which cryptozoology is concerned.* Cryptozoology 5: 1-26

5. ANON, 1972. *Many have seen big monster.* Mainichi Daily News. 1.2.72

6. DOWNES, J. 1995. in SHUKER, K.P.N, 1995 ibid., p.23

7. YOUNG, E and ROSENBLATT, R. 1994 *What's new in New Guinea.* Fortean Times 78:46-47

8. A few unpublished estimates put Deinosuchus at 10-11m (McHENRY, C. pers. comm. 1996) whereas 15m is the figure found in most texts.

9. SHUKER, K.P.N. 1995 ibid. p.24

10. SHUKER, K.P.N. 1995 *New Britain's Lake Monster.* Fortean Times 82: 38-39.

11. LeBLOND, P. 1994. *Almost too good to be true?* The Newsletter of the BC Scientific Cryptozoology Club 19:2

12. Illustration by Lisa Peach on p.39 of ref. 10.

13. p.38 of ref. 10 'He said it was about 33 ft. long'.

14. NAISH, D.W. In press. *Analysing video footage purporting to show the 'migo' a lake monster from Lake Dakataua, New Britain.* The Cryptozoology Review.

15. ROESCH B. pers. comm. 1996.

16. WILLISTON, S.W. 1902. *Notes on some new or little known extinct reptiles.* Bulletin of the Kansas University of Science. 1: 247-254.

17. DOWNES, J. pers. comm. 1996

18. MOORE, J. pers. comm. 1996.

19. pp. 51-52 of STEEL, R. 1989. *Crocodiles*. Christopher
Helm, (London).

20. ALLEN, G.R. 1974. *The marine crocodile (Crocodylus
porosus), from Ponape, eastern Caroline Islands, with notes
on food habits of crocodiles from the Palau Archipelago.*
Copeia 1974 (2):553.

21. MOLNAR, R.E. 1979. *Crocodylus porosus from the Pliocene
Allingham Formation of North Queensland. Results of the Ray
E.Lemley expeditions, Part 5.* Memoirs of the Queensland
Museum 19: 357-365.

22. STEEL, R. 1989. ibid. p.54.

23. NEILL, W.T. 1971. *The Last of the Ruling Reptiles.*
Columbia University Press (New York).

* * * * *

A Genuinely Surrealchemical view of the Lake Dakataua phenomenon

by Jonathan Downes

Regular readers of *Animals & Men* and our other writings will know that many of us at the CFZ are devotees of the archly absurd discipline of Surrealchemy as propounded by its high priest Tony 'Doc' Shiels. During the research for my book *The Owlman and Others* (1997), I read widely on the subject of surrealism after having been instructed to do so by the good doctor. In a book called *Surrealism* written by Patrick Waldberg I came across the following map. It was published in a magazine called *Varietes* in 1929. It is a surrealist map of the world showing each country graded in size according to its importance in the eye of the artist who drew it rather than its actual geographical size.

Devotees and case workers alike will not be surprised to see that the otherwise obscure island of New Britain is situated firmly in the centre of the world, with the Williamuez Peninsula containing Lake Dakataua jutting out proudly and phallically towards the rest of the multiverse. From a surrealchemical viewpoint, therefore, it seems that the migo is more akin to the Midgard Serpent than either a Salty Croc or an aberrant archaeocete. Perhaps she was nothing but a sea head after all.

There she blows....

A MUSICAL INTERLUDE.

EDITOR'S NOTE: Some time ago I was vaguely intending to
record a compact disc of cryptozoological material. As you
may or may not know I am a musician, and at the time my band
'The Amphibians from Outer Space' had just released a CD
('The Case - STP 1995). Dave (the Drummer) and I had just
made a ridiculous public performance at the 1996
Unconvention with Father Lionel Fanthorpe (available, I
believe on chrome tape through Fortean Times), and in the
words of every great entrepreneur it seemed like a good idea
at the time. We had intended to produce a collection
including the 'Ogopogo Funny Foxtrot' (and we had even half
persuaded Karl Shuker to appear on the dance mix delivering
a lecture about the monster of Lake Okanagan). We were also
going to sing a ditty about mermaids, and another about sea
serpents. Both Lionel Fanthorpe and Doc Shiels had expressed
a vague interest in appearing and if the band hadn't split
just before Alison and I did the same thing the project
might have reached fruition.

We wanted some more songs for the compilation and against my
better judgement I telephoned Jan Williams to see if she
knew any songs that could be included. Three days later we
received the following letter:

Dear Jon,

On the telephone the other day you asked me if I could think of any cryptozoological folk
songs. I said 'no' at the time but I had completely forgotten the enclosed ditty which we
picked up at a car boot sale in Ulan Bator. Sadly, the original yellowing parchment crumbled
to dust shortly after I had removed it from a box at the back of the garage. Luckily, however I
had time to transcribe the song (simultaneously translating it from Khalka Mongolian into
English). Enclosed is what I believe to be the only existing copy of the Mongolian lament
"Allergorhai-horhai Batmounkh karakoram" which loosely translates as "Don't mess with the
Deathworm!"

Hoping that this finds you saner than it leaves me,

Best wishes,

Jan.

EDITOR'S NOTE: The fabled Mongolian Death Worm is perhaps the most intriguing contemporary cryptid with multitudinous and bizarre explanations being proffered for its identity.

The fact that such a rare cultural artifact as this song has surfaced lends credence to the existence of a cryptid that some more sober cryptozoologists (myself included) have tended to dismiss as a bizarre and unlikely traveller's tale.

Jan Williams should be commended for her scholarship and for this invaluable contribution to the search for mystery creatures!

* * * * *

Don't Mess With The Deathworm

In a desert in Mongolia
Lives a monster dark and grim
He is loathsome and he's slimy
And his eyes are very dim
Unlike other gruesome creatures
He has no redeeming features
And you'd be advised to steer
Well clear of him

Don't mess with the Deathworm
For his vibes are not benign
Don't mess with the Deathworm
He's a living power line

Hidden deep within the sandscape
There's a Deathworm lying low
Getting ready, incognito,
To put on a shocking show
Should you dig for buried treasure
Or build sandcastles for pleasure
You will find your body
Suddenly aglow

Don't mess with the Deathworm
For his vibes are not benign
Don't mess with the Deathworm
He's a living power line

Don't go looking for the Deathworm
With a long metallic pole
Don't go searching in a sand dune
For the Deathworm's hidey hole
For the Deathworm will be waiting
Keen to do some generating
And finding him will be
Your final goal

Don't mess with the Deathworm
For his vibes are not benign
Don't mess with the Deathworm
He's a living power line.

Don't Mess With The Deathworm

In a des-ert in Mon-gol-ia lives a mon-ster dark and grim. He is loath-some and he's sli-my and his eyes are ve-ry dim. Un-like oth-er grue-some creat-ures he has no re-deem-ing feat-ures And you'd be ad-vised to steer well clear of him.

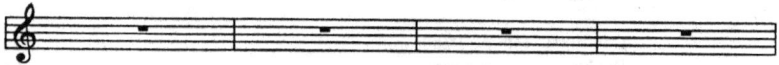

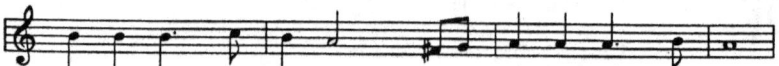

Don't mess with the Death worm For his vibes are not be-nign Don't mess with the Death worm He's a liv-ing pow-er line.

PETSEARCH: FANTASTIC FELIDS AND UFO'S

by Joan Amos

EDITOR's NOTE: The CFZ Yearbook has always been intended as
a forum for the discussion of unorthodox scientific theory.

Joan Amos is a veteran UFO investigator and extremely well
known in her field. She has been a friend of mine for
several years. This article is a distillation of some of her
theories taken from a large volume of her writings.

We would welcome submissions from interested readers -
either for the 1998 Yearbook, or for Animals & Men which
comment upon Mrs Amos' theories and ideas.

* * * * *

I live in a cottage on the western edge of Dartmoor, and in 1978 something happened that changed my life. It was twenty past six on a cold morning in April. I was in a ward of our local hospital with four other patients. We were waiting for our morning cup of tea. The sun was already up and the sky was blue and the window gave us a good view up over the moors. Suddenly we were all very excited to see a halo of bright light hovering over the small market town of Tavistock.

In the middle of the halo of light was a gun-metal grey, saucer shaped craft. It stayed in the same position in front of us, at a low altitude, for six minutes before banking and disappearing from sight beyond the hills. I was rooted to the spot and couldn't believe what I had seen! That was nineteen years ago.

Since then I have read and collected every piece of information that I could find on the subject.

Investigation.

In 1982 I read a book called 'Alien Animals' by Janet and Colin Bord. It had been especially

Dartmoor

recommended in the Flying Saucer Review. I was particularly interested because we have had reports of alien big cats in our area and at the time farmers in our village were losing sheep and lambs.

Mr Dave Nicholls, a UFO contact of mine in Camborne told me of a large yellow cat that he had seen drinking in the river and that he had reported it to Di Francis, a naturalist who was then writing her book 'Cat Country' about the British mystery cats.

I wrote to her and started to send her all my reports. I began to keep a scrap book of these big cat reports and when her book was finally published I read it avidly. I realised that Di Francis was of the opinion that these were an indigenous species of British Big Cat. I was not so sure and told her.

For some time I had good reasons for believing that there is a connection between the big cats and sightings of UFOs.

The first case which I had investigated personally that led me to these conclusions was in March 1982 when a A C.B Radio user was on Dartmoor one night talking to a lady down in Cornwall. Seeing a light behind him and thinking that it was a car, he turned and what he saw frightened him so much that he ran back to his vehicle.

About forty feet in the air was a flying object shaped like an egg with the bottom cut off shining a huge beam of light onto a pony which was rearing up and whining with fright. The bell shaped craft appeared to have a fin on one side and underneath was a ring of small lights which appeared to be turning around. The craft was completely silent.

Within a week of this incident a strange cat turned up at a farm about a mile away. According to the farmer it was a very weird animal which also had a strange effect upon the farm dog. It was a faithful sheep dog which went everywhere with its master, but on this occasion it his under the table and refused to come out. It was the first time that it had ever behaved like this.

The farmer showed me where the beast had leapt over the wall and I took photographs of the large paw marks to show to Di Francis.

The farmer described the creature as "having a snout like a pig, wet and quivering and moving from side to side". When his powerful torch shone into its eyes there was no reflection at all. The head was like a colt and the face was long with pointed ears turning forward. It had a large body with a drooping stomach and a tail like that of a greyhound curved between its legs. When Mr Knowles, the farmer, told his elderly father, the old man insisted:

"It's witchcraft - it's witchcraft".

The moor folk are very superstitious and with good reason.

Research.

This incident gave me much food for thought. I decided to look through my older copies of the Flying Saucer Review.

I found an interesting case in Vol 25 no:2 pp.24. A Canadian schoolgirl had been abducted aboard a transparent UFO. She said that she was able to pass her hands through everything aboard EXCEPT for a large domestic cat. When she asked why there was a cat aboard the UFO she was told that:

"They were growing it, then it would be returned".

In Volume 26 #1 (Spring 1980), Charles Bowen wrote an article hypothesising that mystery animals were involved closely with UFOs. He states:

"We should be prepared to accept that UFOs and alien animals are inextricably linked".

In the same article he described how a alsatian dog cringed frightened in the corner when a 'puma' with a foul smell made nocturnal visits to the farm where he lived. These visits were usually preceded by strange lights, from an undefinable source, playing upon the roofs of the farm buildings.

In Vol 15 #5 there is a description of an animal very similar to that given to me by Mr Knowles the Dartmoor farmer.

"It had a brown head, large black eyes, and a nose extraordinarily like that of a pug. Its left ear was pricked, but the other one hung down like it was torn.. Its ribs were a bright pale chestnut turning to a sort of dirty ginger-brown and its hind quarters were darker still".

Di Francis writes that some of the big cats reported from Britain are *"like a combination of five different animals".*

EDITOR's NOTE: In my book 'The Smaller Mystery Carnivores of the Westcountry' (1996) I suggest that these cats are the result of an incredibly complicated series of introgressive hybridisations between a number of different species of wild cat. This would mean that Di Francis' description was the literal truth!

In a book called 'Secrets of the Ages' the late Brinsley Le Poer Trench wrote:

"John Keel (the well known UFOlogist) tells of similar creatures in the United States and that the Red Indians call them Moon Beasts".

The amerindian legends suggest that these beasts were descended from animals ejected from 'moons' which periodically land in the valleys.

There are a long list of sightings both from mainland Britain and from the Isle of Wight, the Isle of Man and even, according to Cornish radio, on the Scilly Isles.

Di Francis wrote to me:

"Not even with trained snipers and the best equipment have the Royal Marines managed to get a clear shot at the beast and the police are running about in circles. I don't think they'll have a snowball's chance of getting it. One of the Marines was heard to say 'that animal knows what we are thinking before we do'.".

EDITOR's NOTE: In Animals & Men #7 we printed the testimony of a Royal Marine Sergeant who claimed that the 1985 hunt on Exmoor was in fact successful and that one or more animals had been shot. He went on to claim that there had been a government cover-up because of the circumstances in which the animals were shot - outside the agreed rules of engagement for the mission. We treated his testimony with some scepticism, but feel that in view of the claims of a conspiracy by both Mrs Amos and Mrs Francis, that interested readers should re-examine this evidence.

In 'Cat Country' Di Francis printed two photographs for comparison purposes. One shows a fully grown black and white domestic cat standing its hind paws, fully stretched out to grab some meat left dangling as bait. The other shows a KITTEN of the larger species in the same position. It is three times the size and what is even stranger is that it looks just like a large tabby cat, rather than any known species of wild felid.

The South West seems to be especially prolific for sightings of both large cats and UFOs. Animals have been reported which appear to be black leopards, pumas and even a family of lynx. This is just in Devon and Cornwall!

My travelling fishmonger has a customer in Bere Alston who places a regular order for nine pounds of fish which he uses to feed the lynxes which are a regular visitor to his garden.

My hairdresser saw the black leopard on Dartmoor. Together we discussed the sighting in an interview for Channel Four Television. After the camera crew had left my cottage at 8 p.m they drove back over Dartmoor towards Totnes. I suggested that it was unwise. *"Rather you than me"*, I said, but they insisted on going.

It was a filthy night with rain and fog. They got lost and were driving around in circles when suddenly out of the fog strode the black cat.

It presented itself to them in front of the headlights of their car before ambling off down the road. The poor girl shrieked and 'freaked out', not believing what she had just seen. When she telephoned me the next day to tell me of her adventure, I believe that she was still shaking!

The Beast of Bodmin.

In the September of 1993 we had a spate of UFO reports. Huge silent craft, and balls of light were seen at low altitudes. There was also a close encounter case followed by an area of disturbance on Dartmoor. Large granite boulders were uprooted and some have disappeared altogether. It was in an inaccessible location with no signs of wheel marks or footprints. It was almost circular, approximately 48 metres (131 feet) wide from north to south, and 35 metres (115 feet) wide from east to west. There was a dip in the centre of the circle from three to four feet in depth. The grass was visibly flattened, boulders weighing more than half a ton have been moved four or five feet. Some of them left only impressions where they had been. Turfs had been ripped up.

This was reported to us by a man who had tramped across Dartmoor for over thirty years and has never seen anything like it before.

Not long after this the 'Beast of Bodmin' made its appearance. Sheep killings were reported night after night. A woman walking her dogs was also attacked. She was knocked out and when she regained consciousness she saw a large creature that appeared to be stalking around her. Only her barking dogs had kept the creature at bay.

EDITOR's NOTE: This episode is discussed more fully in *Goblin Universe #4* and my book *'The Owlman and Others'* (1996).

The media were soon on the scene and photographs were obtained. People still had great difficulty in accepting the reality of the creature. A farmer on Bodmin filmed the creature and photographs appeared in several newspapers.

EDITOR's NOTE: It is perhaps significant that in view of Mrs Amos' hypothesis regarding outsized domestic cats that the Ninestones Farm footage, on analysis appeared to show exactly that! As noted in both *'Smaller Mystery Carnivores of the Westcountry'* and *'The Owlman and Others'*, we were one of the first investigative teams on the moor. The farm cats owned by Mrs Rhodes of Ninestones Farm were undoubtedly extraordinarily big.

I had just been reading *'The Beast of Exmoor'* by Di Francis. In it she expressed her concerns about a government cover up on the subject. On page 139 she refers to film and photographs which had been sent away for processing which had come out blank although test shots on the same reel had come out perfectly. Still photographs taken at the same time which were developed privately showed the cat and its kittens. What disturbed her most was that this had happened in various parts of the country.

EDITOR's NOTE: The phenomenon involving unsuccessful developing of photographs of a fortean nature is a well known one within cryptoinvestigative circles. It is discussed further in 'The Owlman and Others'. As far as the conspiracy theory is concerned I would refer readers to my previous note on the subject.

Cause for Concern.

In the same book Di Francis writes that in her opinion the animals are more likely to be black panthers (melanistic leopards) than pumas, because there are no records of black pumas and most of the animals which have been reported are incontravertably black. She notes that a pair of breeding leopards in the Devon countryside is a 'terrifying thought'.

Some time ago BBC radio reported that a party of twelve schoolboys was missing on Bodmin Moor. Luckily they were all found safe and sound.

I would, however, like to remind you of some UFO cases. There are many cases on record where animals have been seen on these space ships.

Allen Godfrey, a policeman from Todmorden was abducted. On his return he underwent regression therapy. He spoke of seeing a black animal on board the craft but said that he could not tell if it were a cat or a dog.

A Polish farmer who knew nothing of UFOs was taken into a clearing in the woods after having given two small people a lift on his pony and cart. There he saw a strange craft. When invited on board he saw a flock of birds on the floor. They appeared to be stunned. The little men then asked whether they could examine him. He agreed and was then let go after the examination.

The cattle mutilations are another puzzle. For some reason it appears that our animals are being investigated. A report dated 29.12.1992 states:

"The extraterrestrials obviously have a growing interest in earth and our animal life".

Missing Moggies.

A strange case was reported recently. A half grown kitten was missing from home. It was eventually found stranded on a deserted beach three miles from its home. It was very distressed and was in an almost inaccessible place from which it was rescued by some fishermen who have no idea how it could have got there.

On the same weekend as the close encounter case and the sightings in Tavistock a village in

Cornwall lost thirty domestic cats. *(Sunday Independent 26.9.93).* According to a lady from a local animal welfare charity the police were notified but six months later there was still no trace of the missing animals. The lady I spoke to also told me that late one night she went out of her cottage and she is convinced that she heard the blood curdling scream of a big cat. More recently a hundred black cats were missing from the Clevedon area of Avon.

Cat Killings.

This was the last straw! I recently received a letter from a friend of mine in Canada, a fellow UFO investigator and consultant for the Flying Saucer Review. He enclosed a cutting from a Vancouver newspaper which horrified me. The headline was CAT KILLINGS CONTINUE.

"When Pearl Perechudoff discovered her dismembered cat on Tuesday evening the only way she recognised it was by the markings on its tail. After examining the corpse of the four and a half year old pet, Pearl's vet told her:

'the cat had been cut in half by a sharp instrument'.

The cat, called Kobi isn't the first dismembered cat that has been found in the area. Vancouver's Pound Keeper, Vic warren said:

'Many half corpses have been discovered in the past few years. The cats are cut cleanly in half and disembowelled. Blood is never found around the corpse. The back half is left on a lawn to be discovered by a passer by. Warren said:

'Nobody ever finds the front halves. 'They' obviously want us to find the other half of the animal'.

Warren told us that:

'In 1993 there were nine cases and another five last year. The cut-in-half body of a dog was also found last year'."

Pearl said that she did not want this to go unnoticed. She wanted other people to know that their pets may not be safe. She urged residents to bring their cats indoors after dark and said that *'whoever is doing this is pretty bloody sick'.* It has been suggested that these animals were victims of ritual killings by satanists, but my friend tells me that:

"We did an investigation last year into this sort of event and in our opinion it is NOT satanic cult activity. Our findings fitted the cattle mutilation routine and were confirmed by Linda Moulton-Howe and partner Dr. Auschueler".

I opened up an old book to read at random as I often do. It was called 'UFO Magic in Motion' by Arthur Shuttlewood. On Page 65 I read how Sheila Meredith of Birmingham in 1976 reported:

"It was nine p.m and I was going to get my washing in down at the bottom of the garden. I heard some scrabbling sounds in the dirt. I knew that it was some animal, yet I stopped on instinct and looked up at the heavens. I saw a round, silver object about the size of a half crown at arm's length. As I stared at it, wondering at its brilliance it vanished. I went to the foot of the garden to find out what all the commotion was about in the disturbed soil. I saw that it was my pet cat 'Ginger' who had scratched out a big hole hurriedly. He was very upset and agitated.

There was a thick coat of dirt covering him like a black overcoat, just as if he had dug the hole and been trying to hide from the glittering sphere in the sky above. Believe me, he was visibly shaken and it took me ages to calm him and clean him up".

I wonder what 'Ginger' knew that we don't. Excuse me for now, because it is getting dark and I must get my cats in!

* * * * *

EDITOR'S NOTE: There is a definite correlation between many types of fortean phenomena, and whereas there is a much stronger correlation between certain BHM phenomena and UFO sightings there does, indeed, seem to be some strange relationship between ABCs of certain types and UFO phenomena. I have noted as such in *'The Owlman and Others'*.

As Bernie Mace, of the New South Wales Rare Fauna Research Group noted to me in 1990, there is also a definite correlation between sightings of certain mystery big cats and the disappearances, sometimes on a huge scale, of domestic 'moggies'. Whether this is because, as Joan Amos suggests, the UFO's are abducting domestic pets and filling them with growth hormones, prior to releasing them back onto the moors, or whether as most zoologists will affirm, pumas often prey on domestic cats as a food source, we do not know.

All we do know is that, as the articles in this book prove, the fortean multiverse is a very strange and wonderful place, in which, as Alfred Aloysious 'Trader' Horn once said about Africa: *"Nothing would surprise...."*

* * * * *

King Kong with Fay Wray at the top of the Empire
State Building (1933)

"DON'T WORRY LADIES AND GENTLEMEN - THOSE CHAINS ARE MADE OF CHROME STEEL"

'CRYPTOZOOLOGY' AT THE MOVIES

EDITOR's NOTE: Over a year ago we published a letter from Neil Arnold asking for information about films which featured subjects of a broadly cryptozoological nature. We are pleased here to present Michael Playfair's exhaustive chronology of the subject, together with a personal over-view by the man whose query started the whole thing off.

Crypto on Celluloid.
(A personal view)

by Neil Arnold.

If Captain Sinbad had been alive today, the public would have been more aware of the more unusual aspects to the animal kingdom. In his time, swashbuckling trips to far away places were the norm, and expeditions to find mysterious jewels never cost a dime. He would regularly trudge off, sail the seas and scale the land in search of magic. On his way he would bump into the odd troglodyte or the occasional cyclops, but this was nothing unusual for him. A few of his troops would be torn apart by a sabre-toothed tiger and some of his luscious maidens would need to be rescued from a ravening roc, but in those days it was all part of the adventure. Nowadays the expeditions aren't advertised, they cost a lot of money and the terrain is hostile. Then a kraken in the sea was seen as about as unusual as a sparrow in todays treetops. Nowadays the fabulous beasts have faded into mythology.

The great days of film have also faded. The late 1970's and early 1980's were great times to be scared. Censorship wasn't too rigorous and the old films in their bulky cases were lining the dusty shelves of the video shop. Nowadays many of those old films have disappeared like the Dodo. They were fascinating times - you could turn on the television and get an entire show about living dinosaurs, the horror films looked realistic despite having been made with practically no budget, and a can of coke was only 19p!

It is extremely difficult to obtain any sort of paranormal material these days. It is even more difficult to obtain it at a reasonable price. Books and magazines have to be obtained through specialist dealers, and it is almost impossible to purchase decent video material on the subject. Unfortunately the same themes tend to be regurgitated and anything even the slightest bit revealing tends to be sensationalised (i.e The X Files). The occasional TV programme gets shown, but they are always disappointing, and actual feature films on the subject are few and far between. There are, however, a few out there, and some even have a cryptozoological slant. I am not talking about B Movies which feature radioactive lake monsters, but films which actually take the mystery and utilise it.

EDITOR'S NOTE: What's wrong with cheap B Movies about Radioactive Lake Monsters then?

The subject may be glamourised, or even made into a comedy (i.e Bigfoot and the Hendersons), but some films still utilise a well known cryptozoological mystery. I believe, however, that although there are hundreds of movies out there which deal with these themes, they are usually only available these days, second hand at a car boot sale. Many of these films are ridiculous but some are pretty good. Some dismiss the mystery entirely and others twist it beyond all recognition. I hope that this article will shed some light on an almost forgotten treasure trove of material which disappeared with the majority of 1980's X Certificate films. Hopefully, one or two of you out there will poses some of these gems. If you do, or if you come across any of them, please let me know via the editorial address. So many contemporary films glamourise the UFO phenomenon so that it becomes accessible to the masses. Even after the Hollywood hype, however, the truth remains obscure. Was the *'Hound of the Baskervilles'* really a dog? Maybe. Maybe it was a phantom creature, or perhaps an alien big cat. Let's roll the film.

EDITOR'S NOTE: Conan-Doyle was incontravertable, that the hound was a flesh and blood creature owned by a naturalist called Mr. Stapleton. He daubed this terrifying creature with phosphorescent paint in order to duplicate the old Dartmoor folk story of the Yeth Hound.

Of all the cryptozoological phenomena, 'Nessie' and 'Bigfoot' are film staples whilst mothman and similar creatures are seemingly neglected. It is difficult to know where to draw the line here. Readers of *'Animals & Men'* would certainly be interested in stories of over-sized creatures on the rampage, but it sometimes seems as if millions of films on this theme have been released. *'Jaws'* (giant shark), *'Alligator'* (obvious), *'Night of the Lepus'* (psycho bunnies), *'Cujo'* (beserk dog), *'The Swarm'* (bees) and so on ad infinitum. A lot of these films are quite entertaining if a little basic. *'Razorback'*, a film from the early 1980's saw a giant warthog on the loose in the Australian outback, whilst *'Grizzly'* featured a monstrous bear on a spree.

Almost every animal has succumbed to this trend, and nothing seems too ridiculous to use. Squid, pigs, rats, worms, slugs, ants, birds, wolves, frogs and locusts have all emerged as the animal equivalent of Freddy Krueger...

EDITOR'S NOTE: What about my favourite. *'Pirahna 2 - The Flying Killers'* in which someone managed to cross-breed carnivorous characin fishes from the Amazon with killer bees in what has to be the most ridiculous piece of drivel I have ever seen! It is an essential film in anyone's collection.

There are, however, movies with proper cryptozoological connotations. The most famous, yet oddly the most elusive are the three *'Boggy Creek'* movies. These documentary type films emerged in the mid 1970's and early 1980's and were excellent viewing. They came across as a *'Sightings'* (Sky TV series) type programme, but there was a plot which interspersed fact and fiction. They were directed by Charles B Pierce but now seem to have disappeared. They are probably the most famous Bigfoot films to date.

I'll always remember a scene where a guy was fishing and there was a Sasquatch crouched watching him. Chilling? Extremely. They were also informative, however, and fair to the creatures. The three *'Boggy Creek'* movies were:

'*Legend of*'...
'*Return to*'...
and
'*Barbaric Beast of*'...

Many other films have emerged concerning the hairy man-beast and they are not all ancient obscurities. Recently a low budget film called '*The Abominable Snowman*' was made, but it didn't seem to make it to the video shelves. However, the 1970's film industry did pay homage to the hairy man-beast. '*Werewolf and the yeti*' made the proscribed list of 'video nasties' and is actually just a cheap European gore film with shoddy plot twists. The film doesn't really concentrate on the Himalayan man-beast, but is just an excuse for blood and guts. A cult release, '*The Beast*', concerned a Bigfoot type beast roaming the ruined remains of castles and bumping off people. Once again it is a film which appealed more to the censors than to those with an interest in cryptozoology. The makers obviously thought that it would be original to invent a psycho-yeti. Other films to emerge around this time were '*Yeti*' (1977), '*Bigfoot*' (1971), '*Capture of Bigfoot*' (1974), '*Claws*' (Late 1970's - this concerned a mysterious, shaggy beast supposedly a relative of Bigfoot), '*Bog*' (1978 - which concerned a swamp beast), and in 1976, a documentary called '*Mysterious Monsters*'. Directed by Robert Guenette, it concerned all monsters.

More recently gorehounds were treated to '*Night of the Demon*', which was cut before it was released. This film was directed by James Watson, and concerned a group of trainee anthropologists who go in search of a sasquatch which had killed the members of a previous expedition. Of course, their story isn't believed, but the new team of investigators get ripped to shreds by the creature. Unfortunately, this is another movie that hit the 'nasty' list. Although it concentrates on the Bigfoot mystery it is also heavy on the red stuff. Around about the same time '*Shriek of the Mutilated*' emerged. This was another horror movie which used the yeti as an excuse for buckets of blood. Once again, a group of anthropologists go in search of the snow beast but are savaged. However, they soon discover that it is all a fake and that a mysterious cult is behind the killings. Shame!

One of the best yeti movies was the 1960's Hammer Horror movie, called '*Abominable Snowman of the Himalayas*'. In this film Peter Cushing goes in search of the elusive beast and after a few close contacts with the creatures he returns to base saying that they don't exist, in order to protect them. This film is rare in that it not only has a certain mood, but also shows insight and understanding into the mystery. It remains a classic.

What we do need is factual type films. I know that many films involve various monsters but the more based on fact they are the better understanding is gained of the mystery. Through films like those in the 'Godzilla' series we get nothing but a cheap laugh, but through viewing the 'Boggy Creek' films there is a sense of realism and they never detract from their message that there is indeed a mystery out there. '*Bigfoot and the Hendersons*' seemed promising at the start, when a family on a hunting trip run over the mysterious beast with their car. The film was poor, however as it progressed even though it has now spawned a TV series.

Other obscure films also exist - although some of these are so obscure that they might as well not do. Spanish, Greek and Mexican movies have got in on the act with films such as 'Snowbeast, night creature'. Some of these may have done justice to the Bigfoot mystery, but a majority merely use it as another form of axe wielding maniac. Mind you, Scooby-Doo was quick enough to come across the Abominable Snowman and I even have a scratched old record which went with the series.

'Nessie' and her metaphorical friends have had even more coverage on celluloid. In 1981, 'Loch Ness Horror' was released along with an accompanying book. The film unfairly depicted the mysterious creature as a marauding beast!

(What a surprise! Ed)

'Creature from Black Lake' was made in 1978, and 'Cavern Deep' was another obscurity from the same time. Also in 1978 'Night Creatures' came to the surface, but it was two years earlier in 1976 that 'Nessie' became really popular with film makers. 'Legend of Loch Ness' was a documentary type film directed by Richard Martin, whilst Rock Hudson starred in a fake-'Nessie' TV Movie called 'Birth of a Legend'. In the same year, a Disney cartoon, 'Man, Monsters and Mysteries' depicted the Loch Ness Monster in all its splendour. Also in the funky '70s the 'Tomorrow People' hunted the beast in 'Castle of Fear', whilst Tom Baker as Dr. Who had his own theory on the Loch in 'Terror of the Zygons'. More recently 'Nessie' related movies even reached Germany with a film called 'Nessie, Das Verruckteste Monster Der Welt'! The American spoof movie 'Amazon women on the moon' gave 'Nessie' a little coverage but connected it to the mystery of Jack the Ripper.

The first real Loch Ness Monster movie was 'The Secret of the Loch' from the early 1930's which proves that even over sixty years ago the subject was of irresistible fascination to cinema goers.

Other early movies on the subject were 'What a Whopper', which was a comical farce starring Spike Milligan. Even Simon Templar - 'The Saint' encountered the beast in 'The Convenient Monster' (1966). There was another fake creature in 'Stingray- The Loch Ness Monster' and in 'Journey to the bottom of the sea - the secret of the Loch', an enemy submarine was disguised as a monster.

These are some of the more recognised Loch Ness monster films, although tracking down some of the more obscure ones was a task somewhat akin to that of tracking down the monster itself!

Hopefully some readers will be able to unearth a few films, maybe even cartoons or comedies that have eluded both Michael Playfair and me.

EDITOR's NOTE: Despite the amount of work that Neil and Michael have done, I have no doubt of this. An update will be printed, probably in the 1998 Yearbook.

Of course, in the zany cartoon world anything is possible. Just look at *'Family Ness'* with the talking coloured beasts which chat with the local folk. Yet in its way this is a pleasant cartoon which neither dismisses or smears the reputation of its subjects.

Although as a child I was constantly baffled by it, *'Bunyip'* was indeed the oddest, yet most unique crypto cartoon. Let's face it, who else has done anything about this Australian cryptid?

EDITOR'S NOTE: *Home and Away* for one, although I must stress to all concerned readers that I don't actually watch the programme - honest!

In this series the poor creature found its way to earth after its UFO had crashed! Back in those days programme makers weren't too worried about how educational their programmes were. They took their viewers to new places and opened their eyes as well!

Various other crypto related movies have been released over the years. None are as excellent or as accurate as *'Nature of the Beast'*. This film was shown on Channel Four a few years ago and 1 am still kicking myself for not having recorded it. The film was based on the children's book by Janni Howker. It concerns a big cat at loose on the moors of the north of England. In the film a young boy goes out to capture the beast whilst all the local unemployed townspeople fake photographs in order to earn the reward. This is a superb and atmospheric film. On a similar subject are *'Big Cat'* (1949) and *'Big Cat'* (1988) - two films which seem to have disappeared into obscurity.

In *'Q-The winged Serpent'* a leathery creature, reborn through a weird ritual sets up home on the roof of a New York building. The film is still available and is pleasant if somewhat bloody!

EDITOR'S NOTE: I must admit, that having seen this film on a number of occasions whilst in various stages of sobriety, I have to take issue with Neil about the precise definition of the word 'Pleasant', although I will admit that it is mildly entertaining hokum. This film showcased a trait among Hollywood film directors which I find seriously annoying; the assumption that their audience are idiots. OK, most of them probably are, but is there any real need to insult the rest of us? This film supposedly features the Aztec beast icon Quetzacoatl, but the name was shortened to 'Q' because the PR department at whatever studio was responsible for this cinematic feast assumed that its target audience would be put off by the name 'Quetzacoatl'. A similar marketing decision forced the naming of the film 'Beetle Juice' because the powers that be assumed that the public at large could not pronounce the word 'Betelgeuse'.

'Baby - Legend of the lost Dinosaur' is reminiscent of the mystery surrounding Mokele Mbembe. The film concerns a baby creature similar to a Brontosaurus which is found by some kind hearted trekkers. In *'The Beast of Hollow Mountain'* (1956), two cattle ranchers become irritated at the disappearance of their animals, and they go in search of the cause only to

discover a prehistoric beast. The only problem with this sort of film is that you rarely get what you expect. A well plotted film concerning a living dinosaur is a rare thing, and most of them are in fact corny sci-fi featuring unconvincing rubber monsters.

So, there we have it. That lost is about all that I can come up with for now, although I can guarantee at least another fifty or so films, as well as cartoons and adverts that have featured creatures such as the Loch Ness Monster and Bigfoot. It seems that whereas it is easy to churn out a host of films on the subject of over-sized creatures on the rampage, but films on the subject of expeditions in search of creatures like 'Nessie' or the yeti, films about the reappearance of creatures previously thought extinct, and the search for living dinosaurs are much more rare. It is a pity because these would be of so much more interest.

Unfortunately like so many of the cryptids themselves the vast majority of these films will never be seen or heard of again. Who knows - maybe someone will make a film about Mothman? At the time of writing a film about 'Nessie' starring Ted 'Cheers' Danson is about to be released, and we still get the occasional children's cartoon to get a glimpse of a cryptozoological type creature. If you watch 'Watership Down' you realise that phantom animals are in more films than you would otherwise expect, and after all, 'Chewbacca' in 'Star Wars' is about as 'Bigfoot' as you can get!

* * * * *

CHRONICLE OF FILMS OF INTEREST TO CRYPTOZOOLOGY

By Michael Playfair

Films including animals created by experiments in radiation, etc - or growing to giant size as a result thereof (eg, films such as Them! or The Deadly Mantis) - are not included.

1. VOYAGE OF THE ARCTIC (1903) Includes a scene where a captain of a ship despatches with a revolver a sea serpent which has reared up out of the water.

2. TWENTY THOUSAND LEAGUES UNDER THE SEA (1905) At first, the submarine is believed to be a strange sea monster. Includes a giant crab and a giant squid.

3. UNDER THE SEAS (1907) A submarine has crashed and the captain has to fight giant crabs, sea horses and a huge octopus.

4. JOURNEY TO THE MIDDLE OF THE EARTH (1909) First screen version of the classic story; unfortunately no details exist.

5. 20,000 LEAGUES UNDER THE SEA (1916) A true epic (113 minutes) including Verne's classic mysterious island.

6. TRITON (1917) A sea monster falls in love with a girl - who proves unfaithful, so he returns to the sea.

7. THE LOST WORLD (1925) The best version of Conan-Doyle's story. A landmark film including a pteranodon, an ape-man, a brontosaurus, an allosaurus and many other dinosaurs. The brontosaurus is brought back to London and destroys many landmarks before falling off a bridge and swimming out to sea.

8. MYSTERIOUS ISLAND (1929) Submarine search for half human half fish creatures. They actually look like Donald Duck! Includes a fight with a giant octopus.

9. STARK MIND (1929) Explorers in a South American jungle find a huge ape mysteriously chained (shades of De Loy's ape?); also includes a talonned monster.

10. KING KONG (1933) A cryptozoological classic. An exhibition arrives at Skull Island and encounters not only the huge ape, but a Stegosaurus, Brontosaurus, Tyrannosaurus Rex, Elasmosaurus, Pteranodon; and also a giant vulture. Kong is captured and shipped to New York, where he escapes and is killed on top of the Empire State Building.

11. SON OF KONG (1933) The inevitable sequel to King Kong. This time, a little (12-foot) Kong is found on Skull Island. Other encounters are a Styracosaurus, a cave bear, a dragon-like reptile and a giant sea monster.

12. SECRET OF THE LOCH (1934) The first Loch Ness Monster film; it includes some of Irvine's actual footage. The monster is found to be some kind of giant newt.

13. ONE MILLION B.C. (1940) Cavemen fight giant dinosaurs (well, enlarged reptiles) in this prehistoric love story. The reptile footage appears in many other movies over the years.

14. WHITE PONGO (1945) A large white ape is found, considered to be the missing link.

15. UNKNOWN ISLAND (1948) A lost island is found populated with dinosaurs and, interestingly, a giant sloth.

16. MIGHTY JOE YOUNG (1949) A 15-ft tall jungle gorilla is brought back to civilisation. Won an Oscar for its special effects.

17. PREHISTORIC WOMEN (1950) Stone-age women capturing men and fighting winged dragons.

18. LOST CONTINENT (1951) Rescuers searching an island for a crashed rocket discover a freak environment, preserved by uranium fields and containing prehistoric animals.

19. TWO LOST WORLDS (1951) A ship is wrecked on an uncharted island with prehistoric animals - 'stock' footage from 13.

20. BEAST FROM 20,000 FATHOMS (1953) An atomic explosion awakens a dinosaur, which wreaks havoc in New York before being killed.

21. CREATURE FROM THE BLACK LAGOON (1954) A gill-man found in the Amazon falls for the beautiful girl who is with the expedition.

22. GODZILLA, KING OF THE MONSTERS (1954) A 400-foot tall dinosaur - well, a man in a rubber suit - is awakened by an atomic bomb and wreaks havoc in Japan.

23. SNOW CREATURE (1954) The first Abominable Snowman film. The captured creature is sent to Los Angeles; escapes; rampages; and is killed in the sewers. Many of the scenes were filmed in half-light because of the inadequate suit.

24. 20,000 LEAGUES UNDER THE SEA (1954) Includes the famous scene of the fight with a giant squid.

25. MONSTER FROM THE OCEAN FLOOR (1954) A one-eyed giant octopus is pursued by a mini-submarine.

26. CESTA DO PRAVEKU (Voyage to History) (1955) Four boys voyage down an underground river to a land populated by prehistoric monsters. (The special effects were, for their time, unrivalled.)

27. IT CAME FROM BENEATH THE SEA (1955) A giant octopus (with 6 tentacles!) attacks San Francisco before being destroyed by an atomic torpedo. (Early Ray Harryhausen special effects.)

28. REVENGE OF THE CREATURE (1955) The Creature (gill-man) is captured and brought to an oceanarium in Florida. He breaks out and goes on the rampage before returning to the sea.

29. CREATURE WALKS AMONG US (1956) The last in the gill-man series. The creature is captured again, burned, and loses its gills. Under its scales, human flesh and lungs are found. It escapes, awaiting a sequel which never arose.

30. MAN BEAST (1956) An expedition is attacked by yetis and most are killed.

31. BEAST OF HOLLOW MOUNTAIN (1956) A Mexican rancher finds a dinosaur living in a swamp. It leaves its home and then goes on the rampage.

32. ABOMINABLE SNOWMAN (1957) An exhibition goes in search of the Snowman, which is killed. A series of disasters occur; the film ends with Peter Cushing (the 'goodie') confronting another Snowman, which lures him into the night.

33. MONSTER THAT CHALLENGED THE WORLD (1957) An underwater earthquake hatches prehistoric sea-snails, which attack naval bases before being destroyed.

34. LAND UNKNOWN (1957) An Arctic expedition discovers prehistoric monsters.

35. MONSTER OF PIEDRAS BLANCAS (1958) A lagoon monster attacks and kills people; is pursued by the local sheriff; and falls from a lighthouse.

36. BEHEMOTH THE SEA MONSTER (1959) A dinosaur revived by an atomic explosion attacks London before being killed by a radium-filled torpedo.

37. GIANT LEECHES (1959) Giant leeches attack people in the Florida swamps.

38. GIANT GILA MONSTER (1959) A monster appears from nowhere (out-of-place animal?) and feeds on trees and wrecks trains.

39. JOURNEY TO THE CENTRE OF THE EARTH (1959) Verne's classic story given the big budget treatment: an exhibition to the 'Inner Earth' encounters prehistoric animals (the monsters are lizards) and the ruins of Atlantis.

40. BEAST FROM THE HAUNTED CAVE (1959) A gangster and his hoods hide in the mountains. An explosion releases a legendary monster, which kills them. The monster looks like a cross between a cobweb and a cabbage!

41. OH WHAT A WHOPPER! (1960's) A film I would love to have more information on. A comedy Loch Ness Monster film: the locals are busy faking monsters and the real one appears at the end.

42. DINOSAURS! (1960) A caveman and two prehistoric monsters are accidentally unearthed on an isolated tropical island.

43. THE LOST WORLD (1960) Poor remake of the 1925 classic. The prehistoric monsters are again enlarged lizards.

44. CREATURE FROM THE HAUNTED SEA (1961) A gangster invents a sea-monster to cover his shady plans. A real one appears and devours most of the cast.

45. GORGO (1961) A 65-ft tall prehistoric dinosaur is captured in the Irish Sea and put on show in London. His 250-ft mother appears, and destroys London during the rescue of her young.

46. MYSTERIOUS ISLAND (1961) Prisoners escaping in a balloon are washed up on an unknown island populated by giant animals.

47. VOYAGE TO THE BOTTOM OF THE SEA (1961) A submarine attempts to stop the polar ice caps meeting. Includes a fight with a giant octopus.

48. REPTILICUS (1962) A tail of a prehistoric reptile is discovered. It regenerates itself and goes on the rampage.

49. KING KONG vs GODZILLA (1963) The title says it all. Two actors in rubber suits fight it out; neither wins.

50. SLIME PEOPLE (1963) Nuclear tests awaken scaly humanoid monster near Los Angeles and they create a fog over the city. The monsters are eventually destroyed.

51. 7 FACES OF Dr LAO (1964) Includes a scene where a tiny Loch Ness Monster is exposed to water and grows to giant size, chasing a man.

52. SNOW DEMONS (1965) Expedition discovers snowmen who are aliens with sinister motives.

53. AROUND THE WORLD UNDER THE SEA (1966) Includes an attack by a giant Moray eel on a submarine.

54. ISLAND OF THE DINOSAURS (1966) Explorers discover lost cave complete with dinosaurs and cavemen.

55. ONE MILLION YEARS B.C. (1966) Prehistoric monsters again battle it out with cavemen. This time, the effects (by Ray Harryhausen) are superb.

56. KING KONG ESCAPES (1967) King Kong fights a robot version of himself; includes fights with a dinosaur and a sea monster.

57. ISLAND OF THE LOST (1967) Anthropologist and his family shipwrecked on a South Sea island populated by animals extinct elsewhere.

58. LOST CONTINENT (1968) A ship in the Sargasso Sea finds the Spanish Inquisition and strange creatures.

59. WHEN DINOSAURS RULED THE EARTH (1969) Dinosaurs and cavemen again battle it out.

60. VALLEY OF THE GWANGI (1969) Cowboys in Mexico chase a tiny prehistoric horse into a hidden valley where dinosaurs, including pteranodons and Tyrannosaurus Rex, survive. The latter wreaks havoc in the local town. Has excellent Ray Harryhausen effects.

61. THE JOHNSTOWN MONSTER (1970's) Children decide to fake an Irish lough monster to boost the tourist trade; a real monster appears at the end.

62. THE MIGHTY GORCA (1970) A lost African plateau is found compete with a giant gorilla (Gorca) and other monsters.

63. PRIVATE LIFE OF SHERLOCK HOLMES (1970) Holmes and Watson travel to Loch Ness and find the monster is actually a miniature disguised submarine piloted by midgets! (The model actually sunk in Loch Ness.)

64. TROG (1970) An ice-age troglodyte is discovered in a cave and terrorises and is terrorised by the locals and media.

65. OCTAMAN (1971) An aquatic monster with two legs and four arms falls in love, as usual, with her heroine of the film.

66. LEGEND OF BOGGY CREEK (1972) Semi-documentary about the 'Fouke Monster' - a bigfoot-type creature sighted around the town of Fouke.

67. CURSE OF BIGFOOT (1972) Docudrama. Students uncover a dormant beast.

68. GARGOYLES (1972) Living gargoyles found hiding in an American desert.

69. SANNIKOV'S LAND (1973) A 19th century explorer in Alaska finds a warm area heated by a volcano, where a lost tribe exists.

70. SHRIEK OF THE MUTILATED (1974) Students hunt an abominable snowman but the monster turns out to be a decoy.

71. JAWS (1974) A giant (30-ft plus) great white shark terrorises a New England shore community.

72. THE LAND THAT TIME FORGOT (1975) Based on Edgar Burroughs' story, a German submarine discovers an unknown land populated with Tyrannosaurus, Pteranodons and cavemen.

73. GRIZZLY (1976) An 18-ft tall bear terrorises a Georgia State Park before being killed by a forest ranger with a bazooka.

74. AT THE EARTH'S CORE (1976) Another story based on a Burroughs novel; scientists in a drilling machine discover a world populated by dinosaurs - well, puppets, in this case - and pterodactyls (people dressed up and hanging on wires).

75. CREATURE FROM THE BLACK LAKE (1976) Two men venture into the Louisiana swamps and run into Bigfoot.

76. KING KONG (1976) Poor remake, without the magic of the original. Kong is an actor dressed up, although a full-size robot was made for some scenes. Kong fights a huge snake.

77. CLAWS (1977) A huge bear is again on the rampage but here the suggestion is that the creature is the Ku Tu Ka (a kind of bigfoot) of Indian legend.

78. CRATER LAKE MONSTER (1977) From an egg at the bottom of a lake, a hatched dinosaur goes on the rampage.

79. RETURN TO BOGGY CREEK (1977) Sequel to The Legend of Boggy Creek (1972).

80. SNOWBEAST (1977) A snow beast terrorises a ski resort during a winter carnival.

81. THE PEOPLE THAT TIME FORGOT (1977) Sequel to The Land that Time Forgot. A search party finds a survivor and dinosaurs.

82. LAST DINOSAUR (1977) The world's richest man invades a newly-discovered prehistoric world to stalk a Tyrannosaurus Rex.

83. WARLORDS OF ATLANTIS (1978) Victorian scientists in a diving bell are attacked by a dinosaur and discover survivors of Atlantis. A giant octopus attacks the ship.

84. WHERE TIME BEGAN (1978) Based on Vernes' Journey to the Centre of the Earth, with sea serpents, giant turtles and prehistoric dinosaurs. Much more enjoyable than the 1959 film.

85. THE GIANT ALLIGATOR (1979) Set in Africa! A giant alligator terrorises a holiday resort before being dynamited.

86. FROG DREAMING (1980's) A boy believes that a bunyip resides in a lake. It turns out to be an old abandoned digger!

87. BLOOD TIDE (1980) An underwater treasure hunter sets off an explosion, releasing an ancient underwater monster.

88. THE BOOGENS (1981) A silver mine is opened after 70 years, releasing scaly creatures which kill. They are finally entombed with dynamite.

89. MONSTER ISLAND (1981) People shipwrecked on an island with strange creatures.

90. THE LOCH NESS HORROR (1982) I don't know even if this film was finished. It was supposed to be about scientists finding Nessie's egg which is stolen by a poacher; the monster then goes on the rampage, terrorising the locals, even on the land.

91. RAZORBACK (1984) A razorback the size of a rhino goes on the rampage and starts eating people.

92. ICEMAN (1984) A neanderthal man is found frozen in ice (shades of Heuvelmans & Sanderson) and scientists revive him.

93. BABY... SECRET OF THE LOST LEGEND (1985) Brontosauruses are discovered surviving in the swamps of Africa. The adult is killed and the story follows the attempt of a bad scientist to capture its baby, who is on the run with the help of friendly humans.

94. BARBARIC BEAST OF BOGGY CREEK: PART 2 (1985) Actually, part 3. An expedition to find the monster. Last in series.

95. SEA-SERPENT (1986) A captain attempts to clear his name by searching for a sea serpent which has been raised from the deep by an A-bomb. (The monster is depicted by a puppet!)

96. KING KONG LIVES! (1986) Kong apparently survived the 1976 film and a female mate is found for him.

ADDENDUM

1. A Midnight Episode (1899) A man is awakened by a giant bug crawling over him. He attacks and destroys it along with three more.

2. The Miser's Reversion (1914). A film of interest to cryptozoologists not because of its content but because of a still photograph taken from it. This picture had for some time fooled a number of leading North American cryptozoologists into believing it to be a photograph of a captured Bigfoot. Its true identity was finally revealed due to the research of Strange Magazine's Mark Chorvinsky.

3. Bigfoot and the Hendersons (1987). An American family on vacation hit and injure a bigfoot with their car. They take him home where he takes over their lives and wins their friendship. After many adventures they return him to his family in the wild.

4. The Lost World (1992). Another over-long, poor remake of the classic 1925 film which here is moved to Africa. No enlarged lizards this time, but only half adequate special effects and a cute baby Pteranodon which looks like something which has escaped from The Muppets!

5. Loch Ness (1996). The first big budget production featuring the Loch Ness Monster. A sceptical American scientist arrives at the Loch to disprove Nessie's existence. Gradually his views change, and with the help of one special girl, he comes to believe in its existence before eventually meeting the creature.

* * * * *

The plesiosaur and the mouse: an examination of plesiosaur Morphology and Physiology in connection to their Thermodynamics.

by Stuart Leadbetter

Did plesiosaurs possess an internal heat source or by utilising gigantothermy could they exist in cold northern waters?

We first have to be convinced that the organisms which composed the order Plesiosauria really were reptiles and didn't posses any mammalian features.

Skeletal differences between Mammals and Reptiles.

At first sight demonstrating that a fossilised skeleton once belonged to a reptile might seem impossible. We have no traces of readily identifiable reptilian features such as scales or a skin impermeable to water, but instead of relying on the very visable facets of reptilian morphology and physiology for identification Paleanotologists make use of more subtle characteristics.

The skull from a fossilised skeleton is all that is really needed to identify the class to which its owner belonged whilst still alive. The structures which reveal the truth are the teeth. Although canine and incisors are shared by both reptiles and mammals, at the rear of the mammal jaw are to be found teeth with intricate shapes for specific tasks, for instance our molars for grinding or the carnassials of carnivores for slicing meat. On the whole the mammalian suite of teeth is far more complex than those of a reptile. The jaw itself in which the teeth are located can also be used for identification-the lower jaw bones of mammals are always one solid piece of bone, with reptiles the lower jaw bone is constructed of several bones.

One obvious difference between the skulls of reptiles and mammals is the size of the brain case. Mammalian brains fill more of their skull than the brains of reptiles do but if you were to cut the skulls of a mammal and a reptile open and make an examination of thier respective inner ear systems you would find differences there too. In all modern reptiles a single bone called the Stapes is employed in the transmission of sound from eardrum to the inner ear. The Stapes is also to be found in mammalian inner ear systems but is joined by two other small bones called the Incus and the Malleus. These additional bones were formed from elements of the joint between the skull and the lower jaw in the reptilian ancestors of the first mammals.

The spaces within the layout of a skull also impart information. Palaeontologists use one particular space within fossilised skulls to' classify them. This space is to found directly behind the eye sockets and on both sides of the skull. Not only are they visible in fossilised skulls but they are also visible in modern skull construction. The number and arrangment of these spaces is different for every class of animal and they indicate where the biting point of the jaws is to be found and whether the lower jaw can move horizontally.

Turning our attention to the rest of a fossilised skeleton differences can be found there too. If it belonged to a member of the mamalia the lumbar vertebrae would be free of ribs, in reptiles they continue right the way from the upper thorax to the lower portion of the abdomen. In the region of the shoulder, the shoulderblade or scapula of mammals has a spine running down the centre of this bone, in reptiles this spine is absent. Differences can also be found within the area of the pelvis. Here the three bones which make up this structure-the ilium, ischium and pubis-are fused together in mammals and unfused in reptiles.

There are several other skeletal differences between these vertebrates but taken all together the differences visible within the skeleton of a mammal point unequivocably to its homiothermic metabolism: the complex suite of teeth is required to process food rapidly in order to keep the metabolic furnaces satisfied; the lack of lumbar vertebrae is also a sign of this rapid energy consumption: being free of ribs the lower abdomen will have more space to accomodate an extended gut; the sensitive inner ear system and the large brain are required for active hunting as is the spine running along the centre of both spines: this spine is a site for improved muscle attachment. The fused elements of the pelvis function as an aid to improved locomotion.

The lack of such evolutionary innovations in reptiles explains why they are ectothermic. Of the extinct forms and the forms in existence today only the dinosaurs managed to attain any semblance of a heightened metabolism but they were still ectothermic and like the plesiosaurs of the Mesozoic lacked the skeletal refinements of mammals.

Metabolic differences between mammals and reptiles.

Although we are often told that mammals possess their own internal heat source we are seldom told how this heat source is generated - the answer is by cell matabolism.

In order to perform physical movement or carry out any internal function (digestion, blood circulation, breathing ect) all organisms require energy and they gain either part or the whole amount needed by the chemical burning of food stuffs. The exceptions to this are certain species of bacteria and viruses and of course plants.

There are three metabolic pathways on the way to the full utilisation of consumed food stuffs - Glycolysis and the link reaction; the Tricarbolic acid (TCA) cycle and the Electron Transport chain. The substrates which pass along these pathways consist of Fats, Polysaccharides and proteins. The final product of the whole conversion is the energy storing molecule ATP (Adenosine Triphospate).

Mammals have an advantage over reptiles regarding these metabolic pathways and that is their superior oxyen uptake. Glyolysis does not necessarily require the presence of oxygen in order to function but the Tricarbolic acid cycle and Electron Transport chain do. When Glycolysis occurs in situations of restricted oxygen uptake the term 'Anaerobic Respiration' is used. Mammals often engage in strenous activity where anaerobic respiration takes over from the more normal aerobic respiration so they have evolved well-developed cardio-vascular and respiratory systems as well as the ability to improve their oxygen uptake by training. This means that they can prevent anaerobic respiration from materalising during strenuous exercise over longer and longer periods and even when eventually they must submit to it they can adapt to withstand heavier and heavier accumalations of lactic acid in their muscles.

The story with reptiles is very different. They do not possess cardio-vascular respiratory systems as well-developed as those of a mammal, and therefore have no ability for improving their oxygen uptake.

This then is the source of a mammals internal heat. When oxygen is available (and even when it is in short supply) so much is supplied that excess energy released from oxidating foodstuffs and producing ATP is discarded as latent heat. This latent heat if allowed to build up can be detrimental to the functioning of metabolic enzymes so mammals have evolved mechanisms for keeping their internal enviroment at an optimum level. With reptiles their ability to obtain oxygen is so poor that they only have enough to sustain metabolic activity and certainly do not have very much left over as latent heat. This is their reason for relying on solar energy-after absorbing such energy their levels of activity match those of a mammal.

Predatory strategies in Mammals and Reptiles.

One of the two most obvious differences between mammals and reptiles is due to their differing metabolic rates: reptiles require less food less often than mammals to maintain optimum health. Most mammals prefer to feed at least once a day but reptiles can go days or even weeks between meals. This might appear to be a distinct advantage but it does have one major drawback: having such low intial metabolic rates reptiles cannot indulge in any prolonged strenous exercise like chasing prey so instead they have to rely on patience to catch themselves a meal.

Being unburdened by a metabolism constantly demanding calories, reptiles can lie in wait for an unwary animal for days on end if necessary. To aid them in this waiting they are either camouflaged (for instance snakes) or have low silhouettes (snakes again or crocodiles). When a potential meal does come within range reptiles usually attack in the same way - with a rapid lunge. The different reptile orders have evolved their own individual methods of preventing the prey from escaping once caught the jaw joints of the fish eating gharial crocodile are set in such a manner that the biting force operates through the tip of the snout, perfect for catching swift moving fish; the chameleon, apart from having stereoscopic vision and independently moveable eyes, has a tongue with a sticky end which can be extended to a length longer than the animal itself; many snakes possess fangs through which toxins can be injected into prey either disabling or killing it.

By contrast the predatory behaviour in mammals seldom involves lying in wait for hours on end patiently hoping that a prey animal will blunder into it. The constant demand for energy by rapidly metabolising cells compels mammalian predators to actively seek prey. When prey is located it is captured either by the predators speed (for instance the cheetah); strength (lions); stamina (wolves) or team work (hunting dogs).

Skeletal construction, metabolic rate and predatory instincts in Plesiosaurs; Mammalian or Reptilian?

Simply put, Plesiosaurs skeletons display all the classic signs of a reptilian ancestory: simple teeth which are unspecialised - being in Plesiosaurs simple spikes which could intermesh - an efficient fish trap; the lower jaw of all Plesiosaurs always consists of several bones; small brain cases are evident; the jaw joint is not capable of imparting much sideways movement to the lower jaw due to the single temporal opening behind the eye socket; the lumbar vertebrae always have ribs attached; the scapula does not possess a centrally located spine and the elements of the pelvis are unfused.

Some of these details can also provide evidence for identifying the metabolic rates in Plesiosaurs:

(i). The simple teeth, whilst very efficient for catching fish are very little use in the actual physical digestion of such prey. Once caught the fish would be swallowed whole, as such digestion would be slowed down. This would seriously hamper a Plesiosaur if it had a metabolic rate similar to that of a mammal.

(ii). One of the consequences of having a high metabolic rate is that it allows the possessor to have a large brain. Large brains require constant energy supplies for optimum functioning but if this supply is met it can be maintained by the added intellect improving the possessors to obtain food. This added intellect manifests itself in mammals through such things as planning, team work and improvisation. I very much doubt that Plesiosaurs had the mental capacity to indulge in such behaviour.

(iii). As Plesiosaurs possessed ribs on the lumbar region of the spinal column an extended gut would be very difficult to accomodate there.

(iv). The unfused elements of the pelvis and the ridgeless scapula might not at first seem to be a hinderance to Plesiosaurs at propelling themselves at speeds obtained by marine mammals. After all the sheer size of the bones in a Plesiosaur skeleton tell us that the living creature was heavily muscled. But there the similarity ends. Plesiosaurs used these muscles to drive their huge front and rear paddles up and down in a vertical plane - an efficient enough method of propulsion but the speeds obtained would fall far short of the highest speeds attained by todays marine mammals. So it strikes me as very odd that if Plesiosaurs were indeed endothermic they would channel this potential motive energy into propulsive limbs where only part of it could be utilised. This final point not only tells us that Plesiosaurs not only lacked a high rate of metabolism independent of the external enviroment but it also tells us something about their possible predatory instincts. Having no ability for prolonged swimming activity Plesiosaurs probably had to lie in wait for their prey. Here the reason for their long necks becomes clear: instead of the whole of a Plesiosaur's body moving to capture its prey now only the long neck needs to. Further evidence for this can be found upon an examination of the number of neck vertebrae in Plesiosaurs.

The earliest Plesiosaurs of the genus *Plesiosauridae* which have been found in strata dating from the lower Jurassic had 28 neck vertebrae. During the upper Jurassic period this figure had risen to 44 in the *Elasmosauridae* genus and by the end of the Cretacous period some species in the *Elasmosauridae* genus had progressed to 58 vertabrae and in one species (*Elasmosaurus*) the number of vertabrae had risen even higher to 71.

This demonstrates that instead of evolving a higher rate of metabolism in order to capture prey, natural selection favoured the Plesiosaur's long neck as a better solution. The theory of Natural Selection devised by Charles Darwin indicates that in order for mutations to become fixed in the gene-pool of a species these mutations have to improve the ability of its carrier to obtain food or make it more sexually attractive.

The basic premise of natural selection is that the most valuable mutations are the ones to be found within the genomes of animals which produce more offspring then other animals of the same species. By producing more offspring which will carry some some but not all of the valuable mutations, the orginal animal in which the mutations arose is propagating these chance changes in gene sequence.

It is quite posible that another mechanism involved with natural selection is operating at the same time whilst the neck of the plesiosaur became progressively longer over a million years of selection pressure.

The basic idea behind the term 'selection pressure' is that genotype modifications leading to a change in an organism's phenotype are given a directional driving force due to some interaction; through an environmental factor, through competition with fellow species members or in relationships between predator and prey. The interaction which most possibly placed a directional pressure upon the evolution of plesiosaurs is the relationship between predator and prey. During the span of the existence of plesiosaurs on earth, the fishes that co-

-existed with them were probably their main item of prey. From a study of the fossil fish spanning this time period a general improvement of fish swimming mechanics and body form can be discerned.

What was responsible for this improvement? Could it have been due to mutations which improved a fish's ability to obtain food or its sexual prowess? This is quite possible, but there is another possible explanation as to how this came about. Maybe the selection pressure was induced upon the fishes by a predator - the plesiosaur family.

This selection pressure didn't only run in one direction and upon only one kind of organism - fishes. It also had an effect upon the fish's predator - plesiosaurs. As the fish became better adapted to escaping from the plesiosaur's jaws, a selection pressure would be set up to allow the plesiosaurs to redress the balance between predator and prey. The only way that they could do this would be to improve their only mechanism for catching their prey - the long necks. The selection pressure seems to have only operated in one way upon this area of the plesiosaur - to make the necks longer.

Gigantothermy. Is it a possible solution for the supposed presence of plesiosaurs in Loch Ness?

If plesiosaurs were not endothermic would they still be able to exist in cold climates? A solution called Gigantothermy, (more properly called Inertial Homiothermy) has been proposed. It has been suggested that this is the most important factor that could support the existence of these creatures in the waters of the loch. I, however, take the opposite view.

It is quite correct to state that even if a creature is a polikotherm, it will be a functional endotherm if it has a mass of over 2200 lbs (997 kgs). One important factor is often forgotten whilst discussing this matter - the importance of the basic biological concept of the surface area to volume ratio. To explain this properly and to reveal its relevance to the whole area of discussion about the possible existence of plesiosaurs, we must take a journey into the wonderful world of mathematics.

Supposing one has a cube with sides one centimetre long, what would the surface area to volume be for this cube?

Surface area of cube = length x breadth x number of faces of cube.
Volume of cube = lengh x breadth x height.

Surface area of cube = 1cm x 1cm x 6 = 6 cm
Volume of cube = 1cm x 1cm x 1cm = 1cm

To find the surface area to volume ratio you would simply divide the surface area by the volume.

Surface area/volume = 6cm/1cm = 6 or a 6:1 ratio

Now what would the surface area:volume ratio be for a cube with sides 2cm long?

Surface area of cube: 2cm x 2cm x 6 faces = 24 cm
Volume of cube: 2cm x 2cm x 2cm = 8cm

Surface area:Volume ratio - 24cm/8cm = 3 or a 3:1 ratio.

Do you see what has happened? By increasing the length of the sides of the cube the surface area:volume ratio has decreased. You may be thinking 'so what?' The surface area:volume ratio is a very important concept in biology. There are many aspects of plant and animal morphology and physiology which it is used to explain but the one which I think has the most relevance here is its role in the temperature regulation of animals.

Look back to the two cubes. Which do you think has comparatively more surface area than volume? The answer is the cube with the 1cm long sides. We can now leave the concept of two cubes and instead use two animals radically different in size and in their respective surface area to volume ratios.

The two animals for consideration are a mouse and an elephant. Which of the two has the larger surface area to volume ratio? The mouse does. The volume of the mouse is small and it has a small, thin body and thin legs. The elephant on the other side is large with big cylindrical legs. The amount of elephant on the outside (its skin) is small compared to the amount of mass within its body.

These facts have an important bearing on the amount of body heat that both animals are able to retain. The superior surface area to volume ratio of the elephant means that it has more volume (internal mass) to retain heat in than it has surface area (skin) to lose it through.

I am now going to put another question to you - what do a mouse and a plesiosaur have in common? At first you might say 'nothing' but think of the shape of a mouse - a body with several thin projections radiating out from it. remind you of anything? A Plesiosaur perhaps? It has a body with several thin projections (the neck, tail and flippers) radiating out from it.

Now that we have established that a mouse and a plesiosaur do in fact have something in common we now return to the world of mathematics this time taking with us not a cube but three plesiosaurs.

Surface area to volume ratios calculated for three fossil plesiosaurs.

For ease of calculation and to demonstrate an important fact I have calculated surface area to volume ratios for three different sections - neck, tail and body of each plesiosaur and then added them together to give an overall result for the whole animal. Whilst also looking at these calculations readers may notice that using cylinders to work out surface area and volume doesn't exactly mirror the body contours of a true plesiosaur - the neck would taper towards the head, the tail would widen towards the body, and the body would taper towards both the head and tail. As this tapering occured the reduction or addition in surface area and volume would change uniformly rather than haphazardly in each factor. But this isn't what we are really interested in. What we are interested in is the ratio between the two factors (surface area and volume) and this would remain unchanged.

The first fossil plesiosaur under consideration is *Cryptoclidus*, a small member of the genus *Cryptoclididae* found in rocks dating from the late Jurassic. All calculations are to three significant figures.

Cryptoclidus

- length of neck, approximately one metre.
- length of tail approximately one metre.
- length of body approximately two metres.
- diameter of neck approximately 0.2 metres.
- diameter of tail approximately 0.2 metres.
- diameter of body approximately one metre.

Neck and Tail.

Surface area
= 2 pi r x length of neck or tail
= 2 x 3.14 x 0.1 x 1
= 0.63 m

Volume
= Pi r x length of neck or tail
= 3.14 x 0.1 x 0.1 x 1
= 0.03m

Surface area to volume ratio = Surface area/volume
= 0.63/0.03
= 21 or 21:1

Body.

Surface area
= 2 pi r x length of body
= 2 x 3.14 x 0.5 x 2
= 6.28 m

Volume
= Pi r x length of body
= 3.14 x 0.5 x 0.5 x 2
= 1.57 m

Surface area to volume ratio= Surface area/volume
= 6.28/1.57
= 4 or 4:1

Total surface area = surface area of neck + tail + body
= 0.63 + 0.63 + 6.28
= 7.54 m

Total volume = volume of neck + tail + body
= 0.03 + 0.03 + 1.57
= 1.63

Overall surface area to volume ratio= 7.54/1.63
= 4.63 or 4.63:1

The second plesiosaur to be examined is *Muraenosaurus* an early member of the genus *Elasmosauridae*. This Plesiosaur existed around the end of the Jurassic period.

Muraenosaurus

- length of neck, approximately three metres.
- length of tail approximately one metre.
- length of body approximately two metres.
- diameter of neck approximately 0.2 metres.
- diameter of tail approximately 0.2 metres.
- diameter of body approximately 1.25 metres.

Neck.

Surface area

= 2 pi r x length of neck
= 2 x 3.14 x 0.1 x 3
= 1.88 m

Volume
= Pi r x length of neck
= 3.14 x 0.1 x 0.1 x 3
= 0.09m

Surface area to volume ratio = Surface area/volume
= 1.88/0.09
= 20.9 or 20.9:1

Tail.

Surface area = 2 pi r x length of tail
= 2 x 3.14 x 0.1 x 1
= 0.63 m

Volume   = Pi r x length of neck or tail
= 3.14 x 0.1 x 0.1 x 1
= 0.03m

Surface area to volume ratio = Surface area/volume
= 0.63/0.03
= 21 or 21:1

Body.

Surface area
= 2 pi r x length of body
= 2 x 3.14 x 0.63 x 2
= 7.91 m

Volume
= Pi r x length of body
= 3.14 x 0.63 x 0.63 x 2
= 2.49 m

Surface area to volume ratio= Surface area/volume
= 7.91/2.49
= 3.18 or 3.18:1

Total surface area = surface area of neck + tail + body
= 1.88 + 0.63 + 7.91
= 10.42 m

Total volume = volume of neck + tail + body
= 0.09 + 0.03 + 2.49
= 2.61

Overall surface area to volume ratio= 10.42/2.61
= 3.99 or 3.99:1

The final set of figures is for *Elasmosaurus*, one of the very last plesiosaurs to exist before their extinction at the end of the Cretaceous period.

Elasmosaurus

- length of neck, approximately eight metres.
- length of tail approximately two metres.
- length of body approximately four metres.
- diameter of neck approximately 0.15 metres.
- diameter of tail approximately 0.3 metres.
- diameter of body approximately two metres.

Neck.

Surface area
= 2 pi r x length of neck
= 2 x 3.14 x 0.08 x 8
= 4.02 m

Volume   = Pi r x length of neck
= 3.14 x 0.8 x 0.8 x 8
= 0.16m

Surface area to volume ratio = Surface area/volume
= 4.02/0.16
= 25.1 or 25.1:1

Tail.

Surface area = 2 pi r x length of tail
= 2 x 3.14 x 0.15 x 2
= 1.88 m

Volume   = Pi r x length of neck or tail
= 3.14 x 0.15 x 0.15 x 2
= 0.14m

Surface area to volume ratio = Surface area/volume
= 1.88/0.14
= 13.4 or 13.4:1

Body.

Surface area
= 2 pi r x length of body
= 2 x 3.14 x 1 x 4
= 25.1 m

Volume
= Pi r x length of body
= 3.14 x 1 x 1 x 4
= 12.6 m

Surface area to volume ratio= Surface area/volume
= 25.1/12.6
= 1.99 or 1.99:1

Total surface area = surface area of neck + tail + body
= 4.02 + 1.88 + 25.1
= 31 m

Total volume = volume of neck + tail + body
= 0.16 + 0.14 + 12.6
= 12.9m

Overall surface area to volume ratio = 31/12.9
= 2.4 or 2.4:1

From these sets of calculations the following conclusions can be drawn.

1. In each of the three sets of calculations the neck and tail always has a surface area to volume ratio at least five times greater than the surface area to volume ratio of the body.

2. To decrease the surface area to volume ratio of the body either the length and the width have to be increased at the same time or just the width.

3. The best shape for obtaining the lowest possible surface area to volume ratio is a rectangular cube the shape of the three plesiosaur bodies. The worst surface area to volume ratios result from cylindrical shapes - long in length but short in width like those of the necks

and tails of the three plesiosaurs.

4. Even though the estimates of dimensions used in my calculations may be slightly inaccurate I believe that even if completely accurate dimensions were used the results would still show the following : that the plesiosaurs increased in body width in order to lower their surface area to volume ratios. To supplement this they also shortened the length of their tails and thickened them. They were required to do this because as their necks became longer and longer their surface area to volume ratio of this extremity would increase. Of course these changes in dimensions are proportional because as anyone can see even when *Elasmosaurus* shortened its tail length it was still shorter than the tails of the other two plesiosaurs (see table below).

Total Length	Head	Body	Tail
Cryptoclidus			
4m	1m	2m	1m
100%	25%	50%	25%
Moraenosaurus			
6m	3m	2m	1m
100%	50%	33.3%	16.7%
Elasmosaurus			
14m	8m	4m	2m
100%	57.1%	28.6%	14.3%

After reading these conclusions some people might argue that the results were obtained because although I increased the length of the neck I never increased its width. This would be a fair point, but one which is easily dispelled. Look at the surface area to volume ratios obtained for the neck in each of the three calculations. How can we lower the ratio from 25:1 in one case to a ratio in single figures? The answer is straightforward enough. We keep the length the same but increase the width. Then, however a problem arises. How far do we have to go? 300, 400, 500 mm? In fact to reduce the ratio the neck would have to be as wide as the body, but of course if that happened the resulting animal would not be recogniseable as a long necked plesiosaur.

Another point which some people might raise is that if the overall surface area to volume ratios for all three plesiosaurs is in single figures then how can I argue against plesiosaurs in Loch Ness? Again there is a simple answer. As I have shown plesiosaurs show all the classic characteristics of reptiles. In my opinion therefore this means that they could not possibly a have possessed the regulatory temperature mechanisms of mammals (vasoconstriction and vasodilation, sweating). This being the case, although the body of a plesiosaur would be able to retain a lot of heat the tail and neck would not be able to do so. This would be exacerbated by the fact that a plesiosaur would not be able to prevent heat loss from these extremities.

There is one final point that could be raised in defence of the proposal that plesiosaurs are living in Loch Ness. The Atlantic Leatherback Turtle (*Dermochelys c. coriacea*) has been seen in cold northern waters. If this reptile can survive in such conditions why could a plesiosaur not do so?

As we have seen, whereas the body of a plesiosaur has a low surface to volume ratio, the tail and neck have a ratio as much as five times greater. The Leatherback Turtle has no projections. Its body is very compact giving a low surface area to volume ratio. Whereas it can survive in cold northern waters it is unlikely that a plesiosaur would have been able to do so. These turtles, however, also utilise the heat that they gain from the slow and steady exercise of swimming.

Returning to my conclusions. I can say with a certain degree of certainty that there are no plesiosaurs nor indeed any other aquatic reptile in Loch Ness. One solid conclusion can be drawn from the ratios obtained from the necks and tails of the three fossil plesiosaurs. These extremities would prevent them from living in the cold waters of Loch Ness because they would radiate away precious body heat from them.

Many man made examples of such heat dispersion are found all around us. Take for example a device that is used by many of us on winter nights - the radiator. Look at its shape. It is long but thin. It has a large surface area but a small internal volume. This means that the hot water inside has very little space in which to retain its heat but a great deal of surface area through which to lose it.

I rest my case.

* * * * *

AARDVARKS WITH ATTITUDE

Regular readers of *Animals & Men* will have noticed that it has been our policy over recent issues to produce a section containing a number of short pieces providing different viewpoints of a particular animal. In issue ten we covered the Bengal Leopard Cat and its domesticated 'cousin'. In Issue eleven we covered walruses and in issue twelve the Barbary lion.

For reasons best known to ourselves we decided that it was about time that we did a piece on Aardvarks and so four telephone calls later we are proud to present writings by Karl Shuker, Clinton Keeling and some art work by Darren Naish.

THE UNIQUE AARDVARK
by Clinton Keeling.

The Aardvark - which means 'earth pig' - *(Orycteropus afer)*, which can be translated as 'thing from Africa with claws on its feet gives me great pleasure as a species for the simple reason that I'm the lumper par excellence. I just cannot fathom out the state of mind of the splitter, and here we have an animal of which there is just one species, and it's found only in Africa too. Despite their best efforts the splitters of this world have so far been unable to make even a sub species out of it - such as by saying that those to the east of its range have slightly bigger skulls than those out west, which in any case are inclined to be a bit bigger. No, there's just the one sort of Aardvark or Ant Bear as it was called some moons ago - and nothing looks like it. I must admit that this is the straightforward way in which I like my animals; in fact if I were the almighty instead of just his earthly representative I would create everything along those lines...

In fact, though, there has been confusion and disagreement in the past when it comes to classifying it scientifically, as it was long included amongst the Insectivores on account of its diet of invertebrates, such as termites, but then someone decided that its teeth made it unique and so a special order was created for the one species and dubbed *Tubulidentata* - which means 'having tube or pipe like teeth' - but the latest idea has come from one observant soul who having taken a good look at its powerful and somewhat nail-like claws, decreed it to be a primitive forebear of the ungulates or hoofed mammals. If you are classifying purely through diet, though, it must surely be an insectivore, along with such as the hedgehogs and the moles - which most certainly do not exist on purely insect-based diets.

A superb burrower and tunneller, often into hard-baked and stony ground, it is strictly nocturnal and the possessor of excellent smell and hearing, as is evident from its long snout and large, erect ears; the eyes, however are very small and its range of vision very restricted. It is said to have the unusual habit of burying its faeces like a cat. The skin is sparsely haired, but it is said to have a longer and thicker pelage in West Africa. (It is found all across Africa south of the Sahara). So here is your big chance to create a subspecies. I'll even name it for you. Orycteropus afer hirsutus!

I believe that I'm correct in saying that fossilised remains of the species have been found in both Europe and Asia.

The Aardvark is one of those animals - the Cuscus is another, and so is the domestic cat, that is very tenacious of life, and takes a great deal of killing. In fact it has been known to recover from injuries that would have killed far bigger animals.

This species is rarely to be seen in zoological gardens, partly on account of the great difficulty and expense involved in its being dug out by a gang of men, and partly because it is considered by the directors of some places to be a 'poor' exhibit on account of its nocturnal habits. Once established, however, they do well in confinement and I've known some delightfully tame and friendly ones. The only specimens currently in the United Kingdom are three in the Banhams Zoological Gardens in Norfolk, where they were sent by the London Zoological Garden some four or five years ago, during its period of temporary (we hope) insanity when, forgetting that people come into a zoological garden in order to see animals, it deliberately ran its collection down to a dangerously low level - and then experienced salutory shock when visitors complained about the few animals there...

* * * * *

From the collection of Dr Shuker

AARDVARK ANOMALIES

by Dr K.P.N.Shuker.

"Towards evening the Aard Vark issues from the burrow wherein it has lain asleep during the day, proceeds to the plains, and searches for an ant hill in full operation. With its powerful claws it tears a hole in the side of the hill, breaking up the stony walls with perfect ease, and scattering dismay amongst the inmates. As the ants run hither and thither in consternation, their dwelling falling like a city shaken by an earthquake, the author of all this misery flings its slimy tongue amongst them, and sweeps them into its mouth by hundreds. Perhaps the ants have no conception of their great enemy as a fellow creature, but look upon the Aard Vark as we look upon the earthquake, the plague, or any other disturbance of the usual routine of nature".

Reverend J.G.Wood - *Homes Without Hands.*

The Aardvark is best known to many people as that peculiar beast which is inevitably the first to be dealt with in any alphabetically-ordered animal encyclopaedia written in English - but in reality it has far greater claims to zoological fame than that. Indeed, there is very good reason for looking upon this African anomaly as a truly singular creature - in every sense of the word!

Encountering the Earth Pig.

In 1705, scientist Peter Kolbe travelled to Cape Town, South Africa, and spent a number of years there, during which he acquired a substantial knowledge of the area's wildlife, as later documented in his book *Caput Bonae Spei* i.e. a Complete Description of the African Cape of Good Hope (1719). This contained accounts of many animal forms hitherto unknown to science, - one of which was a strange 8 foot long mammal referred to by Boers as aardvark or erdvark, translating as 'earth-pig'. It derived these names from its subterranean domicile, and from its meat - said by the Boers to taste rather like Pork.

Certainly, it was not named for any overall resemblance to a pig, because in truth the aardvark bears scant resemblance to any other, single creature known to humanity. If anything, it is best likened to an animated collection of spare parts, randomly acquired from all manner of different creatures and almost as randomly assembled, to yield a morphological muddle of very uncertain affinity and highly unlikely appearance - as underlined by the following account, penned by Professor P.M.Duncan in Cassell's *Natural History* (1883-89) and describing one of the first aardvarks to be exhibited in England, at London Zoo:

".... in the evening, and sometimes in the morning, when the food is placed in the cage, a long pair of stuck-up ears, looking like those of a gigantic hare with a white skin and little fur, may be seen poked up above the straw, and soon after, a long white muzzle, with small sharp eyes between it and the long ears, comes into view. Then a very far and rather short-bodied animal with a long head and short neck, low fore and large hind-quarters, with a bowed back comes forth. and finally a moderately long fleshy tail is seen. It is very pig-like in the look of its skin, which is light-coloured and has a few hairs on it. Moreover, the snout is somewhat like that of a pig, but the mouth has a small opening only, and to make the difference between the animals decided, out comes a worm-shaped, long tongue covered with mucus ... and as it walks slowly on the flat of its feet and hands to its food, they are seen to be armed with very powerful claws. In Southern Africa whence this animal came, it is rarely seen by ordinary observers as in England, for there it burrows into the earth with its claws, and makes an underground place to live in, and is nocturnal in its habits sleeping by day".

Its subterranean, daylight-shunning lifestyle could explain the aardvark's success in eluding scientific detection prior to Kolbe's arrival in Cape Town. As for its appearance - why, one could surely be forgiven for suspecting this extraordinary creature to be nothing more than a figment of someone's over indulged imagination, with no basis in reality? And sure enough, following the publication of Kolbe's book, this is precisely the response that the aardvark did elicit. and from no less a person than Count Georges Louis Le Clerc de Buffon - probably the 18th Century's foremost zoological authority.

Discounting the aardvark as an impossible composite, Buffon refused to countenance any possibility that the beast was genuine - until 1766. This is when he learned that a fellow naturalist, Peter Simon Pallas, had obtained further details from South Africa regarding the aardvark and had also presented it with that most cherished manifestation of zoological respectability - a scientific name. In 1795 it was renamed by Prof. E. Geoffroy Saint-Hilaire, and so it was as *Orycteropus afer* ('African digging-foot'), the Dark Continent's enigmatic earth-pig finally, if somewhat belatedly, entered the zoological catalogue.

In due course, moreover, scientists discovered that the aardvark existed far beyond the confines of southern Africa. For example, in Ethiopia a shorter-headed form was found - with smaller ears, a longer thinner tail, and a sparser pelage. Known locally as abu-delaf ('Father of Nails'), this northern aardvark was initially deemed to comprise a separate species, and was christened *O.aethiopicus*. Similarly, a russet-furred aardvark with even smaller ears, formerly described in 1906 from Zaire's Ituri Forest by Prof. Einar Lonnberg, was also looked upon first as a species in its own right, and was dubbed *O.eriksonii*.

However, it is now known that the aardvark has a vast (albeit localised) distribution throughout most of Africa south of the Sahara - stretching from this continent's southernmost tip to Senegambia in the west, the Somali Republic in the east, and Eritrea in the north. Consequently, the Ethiopian, Congolese, South-African, and all other modern-day aardvarks

From the collection of Dr. Shuker

are nowadays classed together as a single species, *O.afer.*

In Search of an identity - Unravelling the mystery of milk teeth and tube teeth.

Once science had finally accepted that the aardvark did indeed exist and was not a hoax, it was then posed with the problem of deciding the aardvark's precise nature. What exactly was this bizarre beast? What were its closest relatives amongst other mammals? Its composite appearance lent itself to a number of different possibilities, each of which attracted due consideration and controversy at some stage.

The aardvark's principal diet consists of termites (often termed 'white ants', even though they are quite unrelated to true ants), which it obtains in great quantities by ripping open termite hills with its formidable claws in a somewhat bear-like manner (which has earned it the alternative name of 'ant-bear'). It snares the hapless insects upon its extremely long, vermiform tongue, liberally coated with sticky mucus and able to flick through the narrowest termite corridors within their demolished edifices.

All of this closely compares with the feeding behaviour of the familiar South-American anteaters or vermilinguas ('worm-tongues'), and as the aardvark does bear a vague morphological resemblance to these mammals too, some zoologists stated that it must represent an Old World branch of the anteater family.

This was subsequently dismissed, however, as further studies revealed that the anteaters were a peculiarly New World group, one which had evolved exclusively in South America, during its many million years of isolation as an island continent (an isolation which finally ended with the end of the Pliocene 2 millions years ago, when South America became connected to North America via the emergence from the seas of the Panamanian isthmus).

As for the aardvark's morphological similarities to the anteaters, these were due not to any taxonomic relationship, but rather to its parallel lifestyle, even though aardvarks and anteaters are not closely related to each other, their lifestyles are very similar, so they have evolved into similar forms - a phenomenon known as convergent evolution.

Another theory put forward was that the aardvark was related to the scaly anteaters or pangolins. Zoogeographically, this had better prospects than the anteater suggestion, because Pangolins are indeed found in Africa. Once again, it was later abandoned, because the aardvark's overall anatomy did not substantiate an affinity with pangolins.

Clearly, its insectivorous diet was responsible for leading zoologists hopelessly astray in their search for the aardvark's taxonomic identity - and then a school of thought arose that has totally ignored dietary considerations, only to offer an identity that seemed even more implausible than those already presented! The English mammalogist Dr. Elliot Smith was startled to discover that the aardvark's cranial structure bore notable similarities to that of a fossil beast called *Phenacodus*, belonging to an order of very primitive, long-extinct hoofed mammals (ungulates) called condylarths. Even the aardvark's bear like claws appeared, when closely examined, to be little more than highly specialised hooves, modified for digging.

An ungulate with claws? A clawed hoofed mammal. It seemed a contradiction in terms, yet was by no means unprecedented. A whole family of horse-related clawed ungulates known as chalicotheres once existed, which are thought to have used their claws for digging up roots and other vegetation, and survived to as recently as the mid-Pleistocene (1 million years ago) in Africa, their last stronghold. Moreover, based upon certain reports of Kenya's famous mystery beast the Nandi bear, some zoologists cautiously retain a hope that the Dark Continent may yet reveal the presence of a living chalicothere. South America also once harboured some clawed ungulates, including the notoungulates *Protypotherium* (a interatheriid) and *Homalodotherium*.

Nevertheless, the idea that the aardvark was a primitive hoofed mammal seemed so radical that many scientists were disinclined to accept it; instead, they sought to reveal its identity by more traditional means. One of the greatest zoological authorities of all time was the late 18th century expert, Baron Georges Cuvier, and such were his talents for taxonomy that he confidently claimed that he could discern the precise relationship of any species of mammal merely from an examination of its dentition. *"Show me your teeth and I will tell you what you are,"* are words attributed to him, and his unparalleled success in mammalogical systematics was ample proof that this was no idle boast. Other researchers practised his methods and it was proved without a doubt that the dentition of mammals was a most important key in deciphering their affinities to one another. Inevitably, therefore, it was not long before the aardvark's teeth received close scrutiny, in the hope that they would expose its true nature. In reality, however, all that they did expose was a fresh indication of the aardvark's omnipresent singularity.

No incisors or canine teeth are present, just 18-20 chewing teeth (premolars and molars), but unlike those of all other mammals, the adult aardvark's totally lack enamel crowns, as well as roots. Their internal structure is unique too - instead of containing a pulp cavity, each tooth comprises a series of closely-packed vertical tubes, numbering between 100 and 1500. In short, these are not really teeth at all, but are simply teeth vestiges - the last remnants of normal teeth that have degenerated over millions of years of evolution. As the aardvark became ever more specialised for an insectivorous existence (in which teeth were of minimal importance). Accordingly, it was impossible to deduce anything concerning the aardvark's taxonomic persuasion from its teth, much to the growing despair of scientists worldwide.

A spark of hope momentarily flickered when, in 1890, British Museum mammologist Oldfield Thomas revealed that young aardvarks possess milk teeth, lost by the adults. Perhaps these would offer a clue? As it turned out, each member of the largest, rearmost pair of these milk teeth does possess a crown and root, but otherwise their structure is still the baffling tubiculous arrangement that had become only too familiar to taxonomists so Thomas' expectations were in vain. (*Proceedings of the Royal Society*, January 1890).

The aardvark was utterly unclassifiable. True, it did seem to share some slight similarities with ungulates, but even these did not appear to be significantly strong enough to justify its categorisation within any of the already-established taxonomic orders of ungulates, living or extinct. Only one option remained - to erect a brand new order, with the aardvark as its sole occupant. And this is exactly what was done, even before Oldfield Thomas's disappointment regarding its milk teeth, when in 1872 Prof. Thomas Huxley chose the aardvark's most distinctive feature as the basis for its order's name - *Tubulidentata* ('tube toothed').

Even today, more than 120 years later, this classification still stands, with the aardvark being deemed to represent an early offshoot from those ancestral ungulates the condylarths, an offshoot that has followed a completely independent, highly specialised course of evolution ever since, totally removed from all other ungulate lines.

A Medieval Madagascan and a native New Worlder.

Discussing the aardvark in *Living Mammals of the World* (1955) Ivan T.Sanderson wrote:

"This creature ... stands quite alone in the mammalian tree of life, like a single green leaf caught adventitiously on a spider's thread".

As already seen above, this is indeed true relative to the present day - but it was not always true. Palaeontologists have discovered several fossil species of aardvark, providing our single modern-day form with a lineage dating back millions of years.

The earliest known aardvark is *Myorycteropus africanus*, a Kenyan representative whose remains date from the early-Miocene epoch, around 20 million years ago. By the mid-Miocene, 10 million years later, the first known member of the modern-day aardvark genus Orycteropus appeared. This was *O.mauretanicus*, native to Algeria. During the late-Miocene, Gaudry's aardvark *O. gaudryi* could be found on the Mediterranean island of Samos, and was somewhat similar in form to Africa's living species but was somewhat smaller in size, measuring a mere 3 feet or so in total length. Remains of other extinct *Orycteropus* aardvarks have been discovered in areas as disparate as Africa (inhabited by *O.crassidens* during the Pleistocene, 2 million years ago), southern France (home to *O.deperthi*) and Turkey (*O.potteri*), in Europe and the middle Sivalik beds of northern India. The European and Asian species date from the lower Pliocene, roughly 5 million years ago. Quite evidently then, the aardvark order had a much greater range in earlier ages than it does today.

True to form, however, just like their single living descendant, the fossil aardvarks are not without their taxonomic tribulations and ongoing controversies. Take, for example, the so-called Madagascan aardvark *Plesiorycteropus madagascariensis*, officially described by Fillhol in 1895. Its remains have been disinterred from several deposits on the central plateau of Madagascar as well as in its western and southwestern portions, as documented fully by Dr Bryan Patterson in a detailed review of fossil aardvarks (*Bulletin of the Museum of Comparative Zoology*, Harvard University 1975), and are only about a thousand years old. In other words *Plesiorycteropus* persisted well into historic times.

What makes this particularly important is that *Plesiorycteropus* was notably different from the African aardvark - due to its shorter skull, its more specialised teeth (lacking altogether, in fact, in some specimens), and also (judging from the structure of its limb bones) to the likelyhood that it was capable of much more versatile movements. Indeed *Plesiorycteropus* may not only have been able to dig, but also to jump and even to climb - talents far surpassing those of the rigidly terrestrial *Orycteropus*. All in all *Plesiorycteropus* presents itself as a decidedly 'un-aardvark-like' aardvark - and perhaps there is good reason for this. In 1988, Duke University zoologist, Dr. R.D.E.Macphee aired the view that this unusual species was not an aardvark at all, and should be placed instead within the taxonomic order Insectovora, alongside the shrew, moles, hedgehogs, gymnures, and related families.

Perhaps the most contentious of all aardvark-associated fossil forms, however, is *Tubulodon taylori* - for two different reasons. Allied by some researchers with the aardvark due to the comparable tubular structure of its teeth, its remains date back to the lower-Eocene, just over 50 million years ago, thereby pre-dating *Myorycteropus* and earning for itself the status of the world's earliest known aardvark - if it truly is an aardvark!

What makes this latter issue so controversial is the locality in which *Tubulodon's* remains were discovered. Namely, Wind River, Wyoming - in the U.S.A. No other aardvark fossils have ever been obtained in the New World, but if those of *Tubulodon* are definitely of tubulidentate affinity, their antecedence of all other fossils indicates that this entire order of mammals actually originated here, and not in the Old World after all. If so, then aardvarks presumably migrated into Africa and other Old World continents at some later date by way of the various land-bridges formerly connecting North America to them.

It would be quite a revelation if the Aardvark were eventually shown to be a native New-Worlder - yet another paradox in the history of this most bemusing beast.

Having said that in '*Vertebrate Palaentology and Evolution*' (1988), Dr Robert Carroll, categorised Tubulodon as an *Epoicotheriid*, belonging to a pangolin like group of mammals known as palaeanodonts existing in North America during the Palaeocene-Oligocene epochs, their taxonomic affinities to other mammalian groups are presently controversial.

Aardvarks in Ancient Egypt?

Notwithstanding its more proliferate past, nowadays the aardvark order is represented exclusively by *O.afer*. However, there are some tantalising clues from the realms of iconography and mythology to suggest that, not so very long ago, even this last remaining species had a greater distribution range than it does today - a range that may have extended as far north as Egypt.

As mentioned by Dr. Jonathan Kindon in *East African Mammals, Volume 1* (1971), there is a pre-dynastic Egyptian vase on record that depicts a number of unusual mammals readily identifiable as aardvarks, thereby implying that in early historical times this creature ranged northwards as far as the Mediterranean region. Of course, a perfectly plausible alternative explanation is that they were inspired by descriptions of aardvarks - recounted back home in Egypt by some of that country's great travellers in those long-departed times, travellers who may well have journeyed through the aardvark's tropical African territory beyond the southern limits of the Sahara. However, there is another, equally thought-provoking piece of evidence favouring the existence of aardvarks in Egypt during historical times.

Set (Seth), one of the more significant deities of ancient Egyptian mythology, was usually depicted as a strange-looking creature whose specific zoological identity still has to be conclusively determined by scientists. Its most noticeable features were a long tubular snout, large upstanding ears, and a thick rigidly held tail. As discussed by H.TeVelde in his authoritative work *Seth, God of Confusion* (1967), a vast and varied selection of animals have been considered as likely (and unlikely) candidates for the Set Beast's identity.

These range from the giraffe, jerboa, camel, hyena, jackal and electric elephant-trunk fish (mormyid), to the crocodile, gazelle, hippopotamus, certain snakes and birds, fennec fox, ass and wild boar. Interestingly, following directly on the heels of its formal scientific discovery in 1901, even the okapi was offered up, by Dr. A.Wiedermann (*Orientalistische Literaturzeitung*, 1902), as a possible explanation for this cryptic creature. And more recently, Michigan Scientist, Dr. Michael D.Swords proposed that it may have been some mysterious form of dog still unknown to zoology (*Cryptozoology*, 1985). There is also one other candidate - the aardvark!

The description of the Set Beast's head and tail certainly compares well with this distinctive animal. Moreover, according to Egyptian mythology the Set beast devoured the moon each month while assuming the guise of a large black boar; it was associated with the desert and desert life; and its rough skin was likened to the short-haired hide of an ass.

If briefly spied in moonlight during one of its predominantly nocturnal forays, the aardvark might well seem to resemble a large, dark, pig-like beast; in the more arid expanses of its range, it does inhabit sandy desert-like stretches; and its skin is certainly rough and clothes only with relatively short hair.

All in all, there is more than sufficient accord between the Set Beast and the aardvark to justify giving serious attention to the candidature of *Orycteropus* as a plausible contender for the identity of Set's arcane animal - which in turn lends important support for the likelyhood that the aardvark did indeed exist in Egypt until at least as recently as the era of Ancient Egypt's mighty civilisation.

Ant Bears, Nandi Bears and a Verismilitude of Vermilinguas.

Not only is the aardvark a very sizeable mammalian mystery in itself, it may also be an unintentional participant in an even bigger one.

For countless decades, Western explorers and settlers in eastern Africa have heard tell from natives of a large and reputedly ferocious beast supposedly still unknown to science, and on occasion they have even apparently encountered it themselves. Bristling with all manner of different local names (including chemosit, kerit and gadett), in English this greatly feared mystery beast is most commonly referred to as the Nandi bear, because rumours and stories concerning it are particularly prevalent in Kenya's Nandi region, and it is said by some to resemble a bear.

In *On the Track of Unknown Animals* (1958) Dr. Bernard Heuvelmans carefully analysed a wide selection of Nandi Bear reports, and came to the conclusion that several different, quite unrelated species of animal were collectively responsible for them This was because the beasts described in the reports were far too dissimilar from each other in morphology and behaviour to be feasibly explained by any single animal species - known or unknown.

The principal components in the Nandi Bear saga appear to be all-black male specimens of the bear-like ratel (honey badger) *Mellivora capensis;* very large hyenas, possibly aberrantly-coloured specimens or even the supposedly extinct short-faced hyena *Hyaena brevirostris;* extremely large baboons; the nefarious activities of local witch-doctors; and, just possibly, a surviving species of chalicothere, which, although not carnivorous, would resemble some of the unidentified beasts included within the heterozygous Nandi bear category. Furthermore, judging from an extraordinary encounter documented in Charles T.Stoneham's *Hunting Wild Beasts with Rifle and Camera* (1933), it seems likely that at least one other creature has made an occasional contribution to the Nandi bear melange too.

The scene was one evening at Stoneham's trading station at Sotik, on the edges of Kenya's Lumbwa Reserve, when, after hearing a noise, Stoneham left his hut and gazed through the mist at what seemed to be an approaching animal of undetermined form. Standing upwind of the animal and concealed by the mist and tall grass all around, Stoneham viewed the oncoming beast with great interest, knowing that it was unaware of his presence and wondering what it could be. When it was within a dozen paces of him, the creature paused. Sudenly, the moon appeared from behind the clouds and illuminated it, revealing at last the appearance of Stoneham's unexpected visitor:

"I received a dreadful shock. The creature was like nothing I had ever seen or imagined. It had a huge square head and the snout of a pig; its eyes, two black spots were fixed upon me in an observant stare. Large circular ears, the size of plates, stood up from its head and they

were transparent - *I could see the grass through them. The creature's body was covered with coarse, brown hair, its tail was the size of a tree trunk. There is an ant-eater in Kenya, a survivor of the ages when the cave bear and the wooly rhinoceros roamed the earth, but this beast, though like to that rare species, was not of it".*

After hearing his account, Stoneham's friends informed him that he must have seen a Nandi bear, but he believed that he saw *"...some weird, hybrid ant-eater"*.

Needless to say, and as Heuvelmans recognised in his own coverage of this incident, Stoneham's account can be dismissed as merely a rather dramatic description of an Aardvark. Its nocturnal nature and noticeably shy, retiring nature ensure that the aardvark is rarely seen, hence an unexpected encounter with one of these extraordinary-looking animals may well come as a great shock to anyone not previously acquainted with it.

In any case, true anteaters (or more correctly vermilinguas), are exclusively Neotropical and belong to the order *Edentata* - which makes the following item especially intriguing. An early 20th Century explorer widely known as 'Stany', (in reality the Marquis of Chatteleux) penned a series of travel memoirs, collectively entitled *Loin de Sentiers Battus*, charting his journeys through the Dark Continent and containing some strange cryptozoological claims.

One of the most curious of these appeared in the first volume of these memoirs, published in 1951, in which he referred to a constant four legged companion called Honore, which he supposedly captured alive in Africa. What makes Honore so interesting is that Stany categorically stated that he was a giant anteater *Mymecophaga tridactyla* - a species endemic to South America! Even Stany's description of Honore corresponded perfectly with this New World novelty! Accordingly, Dr Heuvelmans wrote to him for further details, politely questioning his identification of Honore and wondering (if his pet were indeed an anteater) whether he might have purchased Honore from some importer of Brazilian beasts at an African port. However, in his letters of reply to Heuvelmans, dated 17th November 1953 and 9th December 1953 respectively, Stany unequivocally reiterated that Honore was not an aardvark but a bona fide anteater, which he had personally caught in the African wilds.

Nevertheless, as Stany's memoirs are far from reliable on other zoological matters, it seems safe to assume that whatever Honore was and wherever he originated, he was not a native African representative of the vermilinguas.

Of Pigs and Pumpkins (but not forgetting the Dodo).

The burrowing skills of the aardvark are renowned and unrivalled among large mammals. It can readily out dig a ten-strong team of men, and its immediate reaction to danger is to dig downwards with all speed whenever possible, it is hardly surprising that this remarkable creature is extremely difficult to capture. Its burrow can be as long as 9-12 feet, and is usually about 16 inches wide, with a rounded chamber at the innermost end where the aardvark spends

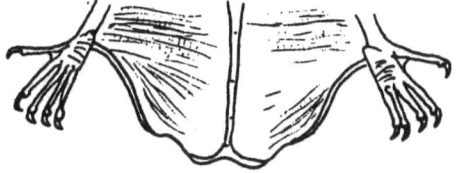

The common slit-faced bat, *Nycteris thebaica*
Picture by Darren Naish

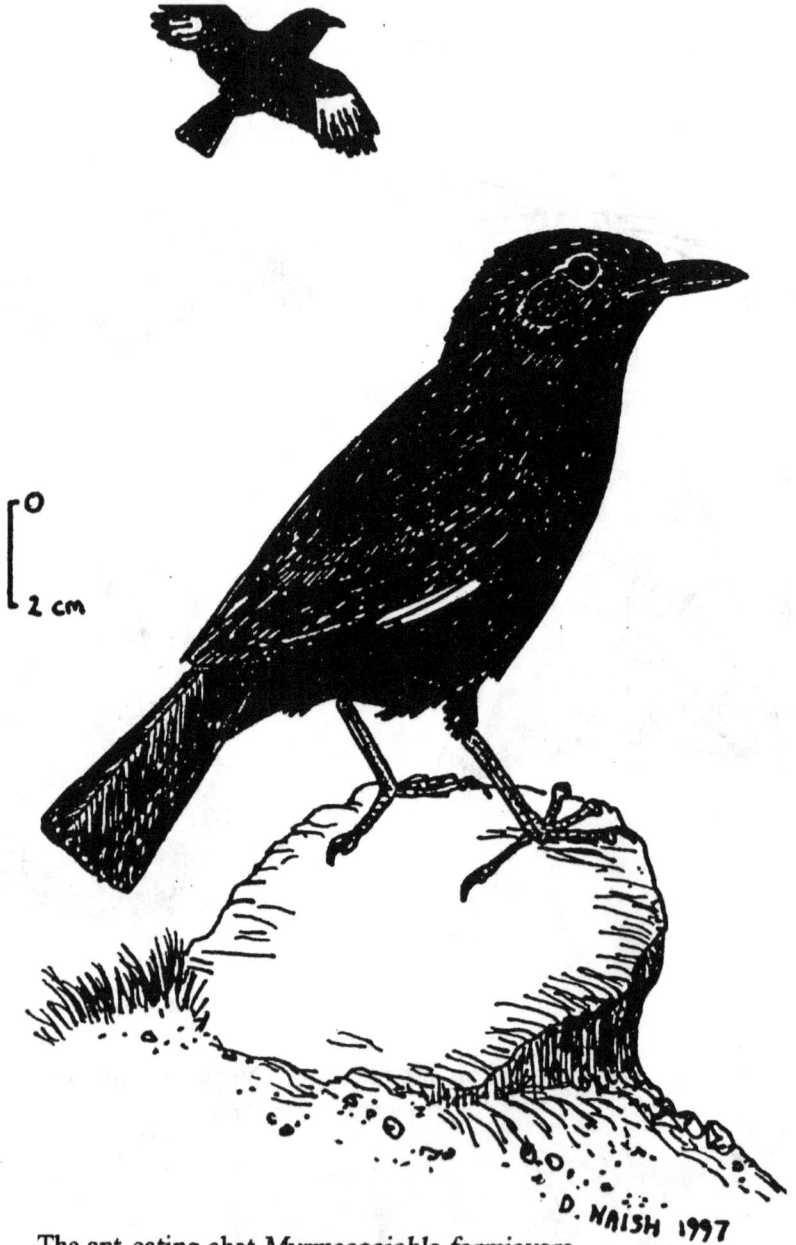

The ant-eating chat *Myrmecocichla formicvora*
Picture by Darren Naish

its days asleep, and where the female gives birth to her single offspring each year in mid-autumn. Every eight days thereafter for the next six months, the female digs a new burrow and shares it with her baby. At the age of six months, the young aardvark begins digging burrows of its own.

Aardvark burrows are in demand not only from aardvarks but also from a surprising range of unrelated creatures, such as assorted owls, pythons, mongooses, lizards, hares and even hyenas. Others include the ant-eating chat *Myrmecocichla formicvora*, (a species of thrush), the common slit-faced bat, *Nycteris thebaica*, and in particular the pulchritudinally-challenged wild pig known as the wart-hog *Phacohoerus aethiopicus*.

EDITOR'S NOTE: Many apologies to Miss Sally Parsons of Eastbourne for this completely unwarranted attack on what the rest of us believe to be a spectacularly good looking creature. She might be interested to know that in a straw poll of a dozen or so warthogs from various wildlife establishments the consensus opinion was that Dr Shuker was no oil painting either!

Whereas many of the previous species listed above often inhabit aardvark burrows on a strictly temporary basis (sometimes as merely momentary hideouts in which to escape from a pursuing predator, or as sanctuaries in which to survive the terrible onslaught of a raging bush-fire), the wart-hog looks upon them as highly desirable long-term residences, with entire families swiftly assuming possession once the burrows' original owners have abandoned them to move elsewhere. Indeed, so popular are these with wart-hogs, that a good way of seeking such structures is to keep a lookout for a wart-hog family - because wherever there is a family of these pigs, the chances are that an aardvark burrow is not far away.

An even more unexpected but equally reliable indicator of abandoned aardvark burrows is the presence of a certain species of gourd-like plant called *Cucumis humifructus* - variously known locally, (and for good reason) as the aardvark pumpkin or aardvark cucumber. Although the aardvark subsists primarily upon termites and other insects, scientists have learnt from various native tribes (after traditionally disbelieving their testimony as folklore!) that in arid areas, where water is scarce, it will consume the juicy underground fruits of this plant if encountered during its burrowing.

The aardvark is a clean animal as far as its own dwelling place is concerned, choosing not to defecate inside its burrows but rather to deposit its droppings just outside, after which it buries them. As these naturally contain the seeds of *C.humifructus*, in due course, the latter germinate, ultimately maturing into adult plants and marking the aardvark's burrows. In fact, it seems that the time spent passing through the aardvark's gut actually increases the seed's fertility, thus providing a noteworthy example of a plant intimately dependant upon the action of a single species of animal for its survival and propagation.

Curiously enough, a markedly similar case to this is one which involved a tree and a very extraordinary bird. The reason for speaking of this case in the past tense is that the bird is now

Warthog *(Phacohoerus aethiopicus)*
Picture by Darren Naish

Dodo (Raphus cucullatus)
From the collection of Dr Shuker

extinct - for it was none other than the famous dodo *(Raphus cucullatus)* of Mauritius.

The case first came to attention in the mid-1970's, when botanists realised that one of Mauritius's native species of tree, *Calvaria major*, was now represented by no more than 13 living specimens - and all of these were over 300 years old, dating back to the very time of the dodo's extinction in the late 17th century. Just a coincidence?

What was particularly curious about this situation was that the last surviving trees still produced perfectly-formed fruit, but the seeds never germinated. Indeed, judging from the synchronicity of dates in relation to the dodo's extinction and the age of the living Calvaria trees, no such seeds had germinated since the extinction of the dodos. American biologist Stanley Temple was greatly intrigued by this mystifying affair, and pondered over the possibility that there was some direct, yet undisclosed link, between the dodo and Calvaria. The seeds of Calvaria were encased within extremely thick, tough shells, and it was this feature that set him on the trail of the mystery's solution. Perhaps the seeds needed some external assistance to escape from their encapsulating shells - and perhaps it was the dodo that had traditionally provided that assistance. But how?

Temple was well-versed in dodo anatomy, which seemed to offer a reasonable answer to that question. The dodo had possessed a very powerful gizzard - a portion of the upper section of the gut in many birds, used for grinding up hard material. If it had included Calvaria fruit in its diet, the tough shells of the seeds would have been broken up whilst passing through its gizzard, thereby permitting the seeds to germinate when defecated or regurgitated by the dodo. It all seemed to fit - but as there were no longer any dodos how could Temple put his theory to the test?

By improvising - which in this instance involved the force-feeding of Calvaria seeds to a congregation of turkeys, which have similar gizzards to that of the dodo. When the various excreted and vomited seeds were collected and planted, the dramatic result was the successful germination of three Calvaria seedlings - which could very well be the first Calvaria to have germinated since the death of the last dodo, in 1681.

From Superstition to Super-Hero.

It is inevitable that any creature as distinctive and mysterious as the aardvark should have inspired its fair share of native folklore and superstition, but like everything else concerning this bizarre beast, one of its most memorable contributions is nothing if not idiosyncratic.

To the witch-doctors of the Hausa tribe from west Africa, the aardvark is a very magical creature known as dabgi, whose nails, heart, and the skin of its brow are of great value to them whilst concocting a charm much sought-after by passionate paramours. Once the aardvark ingredients are ground up with the root of a certain tree, the resulting powder is wrapped up in another piece of pelt, and is secured against the purchaser's chest.

The Dodo and the 'Dodo Tree'
Picture by Darren Naish

According to Hausa lore, the purchaser is now endowed with the useful ability to enter the private room of his lover without needing to seek her father's permission - all that he has to do is to sit upon her roof, and he will instantly pass through it safely and silently. Similarly, a thief owning one of these aardvark amulets needs only to lean upon the wall of the house that he wishes to plunder amid the all-concealing darkness of evening, and the wall will very conveniently part, enabling him to pass unseen and unheard by the house's owner.

And finally: if Africa's insectivorous ant bear makes an implausible ferocious Nandi bear, what can we say about the aardvark as a suave, macho, super-hero?

In fact, since the late 1970's this highly unlikely transformation has been very much a reality - thanks to a quite wonderful comic-book monthly by Dave Sim, called *Cerebus.*

Published by Aardvark-Vanaheim and commencing in 1978, its eponymous star is a laconic tubulidentate of short limbs and even shorter temper, given to saving the world, (not to mention his own skin) at intervals, and all without ever needing a single top-up of termites!

The history of the aardvark is one of unparalleled paradox and perplexity. From milk teeth to tube teeth, medieval Madagascans to ancient Egyptians, ant-bears to Nandi bears, and even from cucumbers to Comic-book heroes, at every turn the aardvark effortlessly underlines its singular singularity, and, above all else, it emphasises its pre-eminent claim to be recognised as the most mysterious mammal alive today!

Select Bibliography.

ANON. *'Hunting the Aardvark; an animal that can dig faster than a Gang of Men'.* Illustrated London News, Vol. 216 (14th January 1950), pp. 54-55.

ARMSTRONG, Peter. 'The Dodo and the Tree'. Geographical Magazine, vol 57 (October 1985) pp 541-543.

BRIDGES, William. *"Africa's Champion Digger".* Animal Kingdom, Vol. 61 (February 1958) pp. 20-22.

COLBERT. Edwin, H. *'Presence of Tubulidentates in the Middle-Siwalik Beds of Northern India'.* American Museum Novitiates No. 604 (30. March 1933). 10pp.

COLBERT Edwin, H. 'Study of Orycteropus gaudryi from the island of Samos'. Bulletin of the American Museum of Natural History, Vol 78. (12th August 1941), pp. 305-351.

HEUVELMANS, Bernard. *'Le probleme de la dentitione de l'Orycterope'.* Bulletin - Museum Royal d'Histoirre Naturelle de Belgique, vol. 15 (August 1939). pp. 1-30.

HEUVELMANS Bernard. *'On the Track of Unknown Animals'.* Rupert Hart-Davis, London 1958.

HOLLMANN, Jeremy. *'A Fruitful Affair',* BBC Wildlife, Vol. 14 (November 1996).

JENSEN, A.S. 'The Sacred Animal of the God Set'. Det Kgl. Danske Videnskabernes Selskab. Biologiske Meddelelser, vol 11, pp. 1-19 (1934).

JEPSEN, G.L. 'Tubulodon taylori, a Wind River Eocene Tubulidentate from Wyoming'. Proceedings of the American Philosophical Society, vol. 71 (August 1932), pp. 255-274.

KINGDON, Jonathan. 'East African Mammals: An atlas of evolution in Africa, Volume 1'. Academic Press, London, 1971.

KNAPPERT, Jan. The Aquarian Guide to African Mythology. Aquarian press, Wellingborough, 1990.

LAMBERTON, C. 'Contribution a la Connassance de la Faune Subfossil de Madagascar. Note XV. Plesiorycteropus madagascariensis. Filhol.' Bulletin de l'Academie Malgache (New Series) Vol 25 pp. 25-53 (1946).

LONNBERG, Einar. 'On a new Orycteropus from Northern Congo, and some remarks on the dentition of the Tubulidentata'. Arkiv for Zoologi. Vol 3, 35pp (1906).

MacINNES, D G. 'Fossil Tubulidentata From East Africa', British Museum (Natural History), Fossil Mammals of Africa, No.10, pp 1-38 (1956).

MONTGOMERY, G. Gene (Ed), The Evolution and Ecology of Armadillos, Sloths and Vermilinguas, Smithsonian Institution Press: Washington 1985.

NAISH, D.W. Pers Comm (re Tubulodon taxonomic status) 14.2.97

NOWAK, Ronald M, Walker's Mammals of the World, Vol 2 (5th edn), Johns Hopkins University Press, London 1991.

PATTERSON, Bryan, 'The Fossil Aardvarks (Mammalia: Tubulidentata) From East Africa', Netherlands Journal of Zoology, vol 25, pp 57-88 (1975).

RAHM, Urs, 'Aardvark', pp 452-458, In: GRZIMEK, Bernard (Ed), Grzimek's Eycyclopedia of Mammals, Vol 4, McGraw-Hill, New York 1991.

ROMER, Alfred S, Vertebrate Paleontology (3rd edn), University of Chicago Press, London 1966.

SAVAGE, Robert J G & LONG, Michael R, Mammal Evolution: An Illustrated Guide. British Museum (Nat Hist). London 1986.

SHOSHANI, J, et al. 'Orycteropus afer', Mammalian Species No.300, 8pp (1988).

SHUKER, Karl P N, In Search of Prehistoric Survivors, Blandford Press, London 1995.

'STANY', Loin des Sentiers Battus, Vol 1, Table Ronde, Paris 1951.

STONEHAM, Charles T, Hunting Wild Beasts with Rifle and Camera, Hutchinson, London 1933.

SWORDS, Michael D, 'On the Possible Identification of the Egyptian Animal-God Set', Cryptozoology, vol 4, pp 15-27 (1985).

TeVELDE, H, Seth, God of Confusion, E J Brill, Leiden 1967.

THOMAS, Oldfield, 'Milk Dentition in Orycteropus', Proceedings of the Royal Society, vol 47 (Jan 1980) pp 246-248.

THOMAS, Oldfield, 'Name of the Aardvark', Proceedings of the Biological Society of Washington, vol 14 (2 Apr 1901), p 24.

WENDT, Herbert, Out of Noah's Ark, Weidenfeld & Nicolson, London 1956.

* * * * *

COMEBACKS

(an idiosyncratic look at the proposed reintroduction of
certain animals, once native and now extinct, into Scotland)

by Tom Anderson

Scottish representative of the Centre for Fortean Zoology
(Is it a bird? Is it a plane? No it's Aberdeen's Mr Entertainment - he sings, he dances, he jumps naked out of a cake!)

1. The Wolf

Fossil remains of *Canis lupus* dating back to the Pleistocene have been found in Norfolk, the Pentland Hills in Scotland and Shandon in Ireland. In Anglo-Saxon times January was traditionally the month for wolf hunting presumably to lessen the predation on their stock in the spring. The lack of cover and attendant winter loss of body weight of the wolf would have aided in this.

In 1427 laws were passed in Scotland as the wolf population was thought a major menace. Apparently this was of little effect as during the late sixteenth century offshore islands were utilised as burial grounds to avoid cadaver predation.

The threat would seem to have been greatly overstated. As is usually the norm in these things, and by the time of the Tudor ascendancy the last English wolf was no more. North Yorkshire and Cumbria vie for the honour of having the last English specimen killed on their soil.

(EDITOR'S NOTE: At least one village in Cornwall also claims this dubious honour).

In Scotland and Ireland, however, more broken country and a widely scattered population coupled with different farming methods meant that the wolf survived until the mid 18th Century.

The folklore and legend end at this point. We all know about the Paris citizenry besieged behind the city walls and Russian nobility being pursued in their troikas by packs of ravenous wolves. Both of these can be interpreted as euphemisms for the downfall of Bourbon and Romanoff. This is heady enough, but when the spectres of repressed sexuality and other hang ups, freudian or otherwise, are introduced, I'll stick with biology.

The last year has seen a sudden upsurge of specialist interest in exotic reintroductions into Scotland. The undoubted success of sea eagles brought from Norway, Red Kites from other sources have no bearing on the feasibility of large mammalian carnivorous or omnivorous additions.

The european wild boar is farmed in Scotland, feeding time is terrible to behold, and their imminent release would not assist the pate market, so in their pens they'll stay.

The ostrich, both blue and black backed varieties is even more prolific in captivity. Fortunately it could not be a re-introduction so we're safe, tho' the Easter egg-rolling season could prove hazardous.

Early in 1995, Professor Gorman of Aberdeen University postulated the Isle of Rhum, off Scotland's west coast, as a likely release area for the wolf. It has a resident red deer population and little else, being roughly the size of Newcastle.

This suggestion, and others from Oxbridge and other sources were refuted by 'The Establishment'. The Scottish Office, The British Deer Society, The National Farmer's Union, and sundry others all voiced their disapproval.

The 'For' argument is that the wolf, having been a previous resident thereby qualifies for re-patriation. It would be a tourist attraction and cull the over populous red deer 'naturally'. The case of Latvia is cited as a type example but is not comparable to the Scottish situation.

The best projections claim between ten and fifteen years in 'ideal' areas such as Glen Affric, or a similar wooded area as the closest date of release. Other, less afforested areas could take fifty years. THAT is the main 'anti' argument.

Scotland's tree cover has been devastated to the extent that Red Deer are now residents of the open hill, not their natural habitat. Neither is it the wolf's. It has been estimated that 50% to 75% of Scotland has to be re-afforested to create a viable environment for the wolf. This won't happen! Further down the line the 'argument against' hypotheses include the argument that they would interbreed with dogs. They have a common ancestor in *Amphicyon*. If this did happen a much more intractible cross breed would emerge. The current North American hybrids which have more than 20% wolf blood are totally unsuitable in a domestic situation.

The argument for it as a tourist attraction is hard to sustain. Edinburgh Zoo has a 'Wolf Wood' stocked by Canadian timber wolves. I've yet to spot one and that's in a fenced enclosure. The 'natural' cull is equally untenable; on the Highland estates, even a hind has monetry value when shot by a 'sportsman', whereas venison tends to lose its appeal when indented by canine incisors. Until such time as the deer population stabilised, based on the annual cull figure, wolves would have to kill 384 red deer per week, minimum.

This would reduce the red deer population to a level consistent with the available habitat.

When the optimum level is attained, do we then start culling the wolves?

2. The Wild Boar.

The European wild boar (Sus scrofa) has never been rife in Britain since it was systematically hunted as sport. Around the year 940 edicts governed its hunting by nobility to the month of November. Re-introductions began with Charles I and continued through to an unconfirmed report from Dorsetshire in the first quarter of the nineteenth century.

Its appeal as a quarry capable of killing its persecutors mitigated against its survival. Mankind, certainly the western european variety engaged in 'field sports' experience some unease when the "pelt" bites back.

In the spring of 1996, one David Hendry (33), intended to fence off a thousand acres of Perthshire woodland, prior to introducing thirty wild pigs. He claims that after three years his clientele would be able to 'crop' around fifty boars annually.

Boar hunting is very popular in Germany and Poland where the animal is still relatively common. Having seen this in action, I have yet to see any pig flushed from cover to cross the guns, approaching anything like a full grown boar.

Glasgow charity Animal Concern claims Mr Hendry's scenario is a *"perversion, not to be accepted in any civilised society"*.

Mr Hendry planned to finance the scheme from the £3 million he received from the sale of a previous enterprise - a funeral business!

3. The Beaver.

The earliest reference to Castor fiber in Britain is in the Welsh laws of Howl Dha, in A.D 940. A major dictat is that:

"The King shall have all Beavers, Martens and Ermines, because from them the borders of the King's garments are made".

The price of a beaver's skin was then 120 pence, whilst a marten was only 24 pence, and that of a wolf, a fox and others 8 pence. This shows that even then the beaver was rare, or possibly the reasons for the rarity.

From the 12th Century, according to Gerald de Barri, the beaver's musk glands, much prized since Roman times as an aphrodisiac, led to its persecution. I quote Barri's dictum on the beavers of the Welsh river Teivi:

"When the beaver discovers that he cannot escape from the attention of the dogs which follow him, he may decide to ransom his body by the sacrifice of the part which by natural instinct he realises they prefer. In the sight of the hunter he castrates himself from which circumstance he has already earned the name of Castor. If by chance the dogs should chase an already castrated individual he has the good sense to run to an elevated spot, lift up his leg and shews the hunter that the object of his pursuit is gone. Thus, therefore, in order to preserve his pelt which is sought after in the West, and the medicinal part which is coveted in the East, although he cannot save himself entire yet by a wonderful instinct and sagacity he tries to avoid the stratagems of his pursuers".

Beloved not only of apocatheries, it's tail was also highly prized, being first roasted and then simmered to 'dispel the evil vapours'.

Its past presence is attested to by place names such as Beverley, literally 'The stream of the Beavers'. Confirmed by bones found by the nearby River Hull.

As with the Reindeer its extinction progressed northwards until its ultimate demise in the early 18th Century.

In 1860 the 'Society for the Acclimatisation of Animals, Birds, Insects, and Vegetables within the United Kingdom', was formed. Its founder and the author of this snappy little title was Francis Buckland, former surgeon in the Life Guards, Inspector of Fisheries and subject of a recent book by Sir Christopher Lever called 'They Dined on Eland'.

In a lecture in November 1860 he proposed as suitable subjects for acclimatisation; Bison, Elk, Eland, Reindeer, Kangaroos, various deer species, Beaver, Wapitis and Yaks.

Reindeer already being present on the Cairngorms, the beaver's reintroduction is imminent subject to agreement and certain conditions being met.

The Canadian and European animals are so alike as to be cast as one species, genus *Castor*; with listed sub-species. Originally found all over Europe east of the Pyrenees and much of central and northern Asia it survives only in protected areas, along the German Elbe, Poland, Scandinavia, Russia and below the Roclanes in France.

Recently reintroduced into Switzerland where it had been extinct since 1820 its summer diet is meat, shrubs and water lily roots. In winter in exists on fresh bark. The Canadian is more prevalent in zoos as it breeds more successfully in captivity than the European.

It is a large animal by rodent standards. 1.3 metres long including a 30cm tail. It has a unique toe which serves as a toothpick and a 15cm broad tail which doubles as a rudder and a paddle. Normally jet black, a light cream specimen with a white tail reported in the US would appear to be an aberration.

Superficially a monotreme, closer study reveals the standard number of orifices plus the oil preening glands and the castorum. The latter is both a territorial marker and its own undoing when used as trap bait. It is also reportedly useful as a dressing on flesh wounds. The kits are born in April in litters of up to six, the females apparently practise infanticide on half their young during the two weeks following birth when the males move out. Reports of freshly killed kits floating in the lodge pond are on record.

Being a true wilderness animal, Scotland'would seem a more suitable habitat than, say, East Anglia. Their bark stripping tendencies could prove a problem and some form of rotational re-afforestation might be required in the long term. Having a more static lifestyle than the others, it would also alter the landscape more visibly than perhaps is desireable.

REFERENCES

FREETHY, R. *Man and Beast*. (Blandford)
TOPSELL, E. *The Historie of the Foure-Footed Beastes*. (London, 1607. Jaggard).

N.B. On February 7th 1996 a beaver was killed by a car at Glenfalloch in Ayrshire. A second was found exhausted by children playing in the snow. Reports of two adults and one cub having been seen are likely to refer to these animals.

Last year (1995) a number escaped from a collection near Loch Lomond but, as the colony is presently hibernating, any shortages will not be missed 'till April.

(EDITOR'S NOTE: These articles were submitted to us between November 1995 and March 1996).

Following a succesful project in Norway 25 years ago, Scottish National Heritage intend to re-introduce the beaver as soon as is practical.

Further source material:

RUSSELL Osborne. *Journal of a Trapper*. (Oregon Historical Society 1955 Ed.).

4. The Lynx *(Felis lynx)*.

The Lynx takes its name from Linceus, who, according to Greek Mythology, was gifted with sight so acute he could see through opaque objects.

Modern experiments have proved that a lynx can distinguish a mouse at seventy five metres, a rabbit at three hundred and a roebuck at five hundred metres. The European lynx differs from the North American bobcat *(Lynx rufus)* by its colouration, five centimetre ear tufts and long cheek ringes.

Despite its stocky build it can leap almost three metres into a tree and can pounce extremely quickly. Its main prey being birds, reptiles, amphibians, squirrels, grouse and occasionally deer. The bobcat is so dependant on hares that its population fluctuates in accordance with that of its prey.

Most at risk is the Spanish lynx *(F. pardina)* which formerly inhabited Italy, Sicily, Sardinia, France, and the eastern Mediterranean. Now restricted to four small enclaves on mainland Spain its last stronghold is the Donana National Park where it is carefully monitored. It has been considered at more risk than the Asian tiger for the last twenty-five years. A combination of persecution, habitat destruction, and the young's first year dependance on its parents have all contributed to its current status in the 'red book'.

A programme of reinforcement and reintroduction of the wolf, bear, lynx and wolverine back into mainland Europe was recently launched by the WWF.

A laudable project, aimed mainly at regions like the Carpathians and the Sierra Morenas, ideally suited for large tracts of state owned wilderness where a niche exists for a reintroduced predator. The situation radically differs in Britain where one major state landowner, the Forestry Commission, specialises in vast tracts of coniferous monoculture, barren of prey and predator. The main justification for bringing back the wolf was as a natural culler of the now excessive red deer herds.

Stalkers could achieve the same result in one/two seasons; moves are underway as I write. Our native wildcat fills the niche for a predator felid, a second would be superfluous, if not hazardous to the resident. Anyone wanting to release bears here would have to stand the cost of a high fence encircling most of Inverness-shire.

Probably the least desireable candidate for naturalisation if the Wolverine. Anything capable of bringing down a young elk and having the tenacity to 'tree' a human for two days and nights is the stuff of a nightmare. Reports of its presence in Wales and the west country must be the worst news for years. Given its ability to evade, capture and withstand persecution in environments climatically much more hostile than our own, we can only hope that the outbreak of mange currently decimating fox populations thereabouts lays it low.

Placement of species has always been contentious. If not downright dangerous to the original inhabitants, at least misguided, with little regard to future implications. Our island is of a scale incapable of withstanding even minor mammalian imbalance eg - mink, coypu. Both species were misplaced; the former, alas, still wreaking havoc on native waterfowl and, according to research over twenty years, displacing the otter from its natural habitat.

Muskrat, grey squirrel, porcupine, asian deer, (various) wallabies, polecat-ferrets etc. All have experienced varying degrees of success and, almost without exception, have not enriched our countryside sufficiently to justify their presence.

At present only three reintroduced predatory species of any substance are likely to succeed. Osprey, White-Tailed Sea Eagle. Red Kite, all avian, two piscine in diet, one opportunistic feeder. Is this as far as we dare to go?

REFERENCES.

LEVER, Sir C. Naturalised animals of the British Isles. (Paladin, London 1979).
SALVADORI, F.B & FLORIO, P.L. Wildlife in Danger. (1978 English Translation).
Red Data Book (IUCN Morges, France).
Press Release 16th February 1996 by Magnus Sylven, WWF European Director.

* * * * *

Were aquatic pre-humans the first vertebrates to enter the land?

by Francois de Sarre

President of the CERBI (Study and Research Centre for Initial Bipedalism), Nice, France.

Member of the SEI (Societas Europea Ichythologorum), Frankfurt-am-Main, Germany.

French correspondent for the Centre for Fortean Zoology.

When I was a young zoology student, I assisted at one of my first lectures at the University. I remember that it dealt with the classification of the vertebrates, and especially with the origins of the first land dwellers amongst them.

In one corner of the auditorium there was a splendid placard hanging that referred to this topic by depicting an ancient coelacanth above a representation of *Ichthyostega*, an early amphibian which was allegedly the remote ancestor of newts, frogs .... and ourselves!

The Professor was discussing the filiation within vertebrates. He made the tetrapods (i.e four legged animals with prevailing terrestrial habits) descend from an odd fish that seemed to be a coelacanth. "Sure", he said, "*it was not the living species now known as Latimeria that once climbed onto land but the ancestor that we have in common with all tetrapods : Eusthenopteron, from the Devonian period, about 400 million years ago*".

When the lecture was over I examined the illustration of the coelacanth and its amphibian progeniture. I could not tell why, but I wanted to reverse the schematic so that the coelacanth would appear as the 'fully aquatic' version of the *Icthyostega*! To me it looked as if the evolutive process was working towards the classical design of the fish, rather than the diametric opposite!

I expressed my problem to the lecturer who was standing besides me and asked him why my theory would not work. He answered that it would not fit chronologically because whereas the fish fossils had been found in the Devonian layers, the *Icthyostega* was known from the Carboniferous era. The *Icthyostega* was therefore only 350 million years older hand had appeared later on the scene of evolution. I replied thoughtfully:

"*But indeed sir, might not the geological archive be still incomplete, and could not new discoveries in the future change our point of view?*"

"*Don't talk nonsense!*" he said to me. "*It is a proven fact that land vertebrates evolved from fish!*"

Within the Fable of an Earth-conquering fish.

Many years have passed by since then, and my chosen domain has indeed become the study of the fish species of the Mediterranean. Since 1968 I have published several papers on this topic.

I have not forgotten my dilemma about the coelacanth. The further that I have advanced with my own reflections of vertebrate natural history, the more I have become convinced that fish have a terrestrial origin. This means that their ancestors were four legged and air breathing animals that lived both on land and in water as do our present day amphibians.

Contemplating the manyfold fish species of the present day, it seems no great exaggeration to extrapolate that not just coelacanths and other groups such as lungfishes and the toad like *Periopthallmus*, but all fishes descended from terrestrial ancestors. This has a precedent in the Ichthyosaur which was a finned and fish shaped reptile, as well as some mammals which according to conventional belief have 'turned back' to live in the sea. If zoology is to continue considering fish as the 'primordial vertebrate', then it is simply a matter of usage.

From recent discoveries on palaentology we might indeed learn, if we ignore our prejudices, that in the remote past lived both the Crossopterygian fishes and amphibians which looked very like them. They were contemporaeneous and lived together!

From a purely structural view it is surely easier to explain how the well formed leg of an amphibian could have developed into a fish fin, than the opposite which indeed takes a leap of faith.

As I began to realise; the old cliche of an evolution from fish to us had to be totally overthrown and had nothing to do with the actual history of the vertebrate groups.

The theory of Initial Bipedalism
or
An encounter with the real facts of evolution.

I was continuing with my reflections when an important event happened in my life. I met Dr. Bernard Heuvelmans, a famous French zoologist who has been justly called the 'father of Cryptozoology' (the science of hidden animals). We had a lot to talk about, and our conversations were not just on the subject of undiscovered large species such as relict hominoids. We also discussed the concepts of vertebrate evolution, and last but not least, the prickly problem of Man's origins.

Bernard Heuvelmans told me about the theory that had been developed in the 1930's by his master, the mammalogist Serge Frechkop, and by the German anatomist Max Westenhofer. This was the theory of Initial Bipedalism.

If we look carefully (see Figure One), at the different stages of development of vertebrate embryos (man, dog, chicken and tortoise), it is surprising to find that we can hardly tell them apart. At a certain age (below) some differences can be noted which correspond to the general evolution of their lineages. Indeed, the human remains a biped, the dog develops into a quadruped, the chicken becomes a winged biped and the tortoise evolves into a carapace-bearing reptile.

The illustration in figure one draws our attention to the fact that the animal embryos are very similar to the human prototype and that in the earliest stages they all exhibit a round, big brain inside their embryonic skull. At this stage they are ALL potential bipeds!

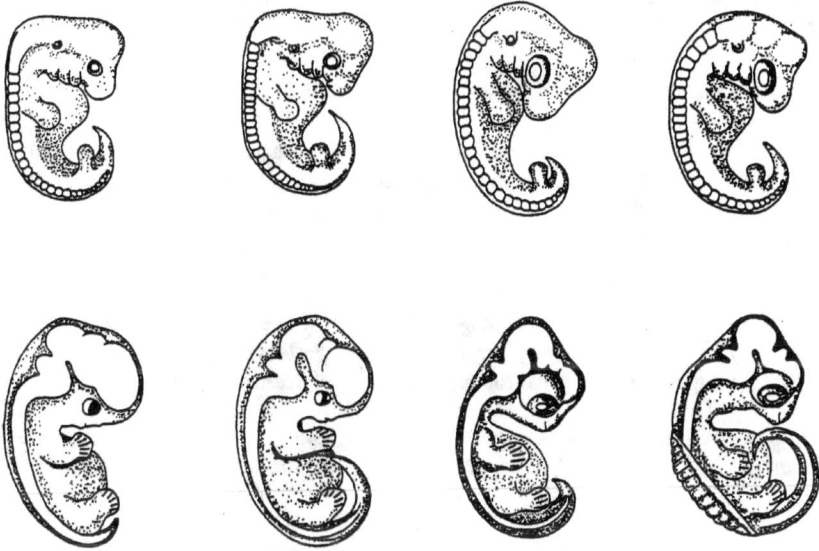

Fig. 1   E M B R Y O N I C   S T A G E S

of

M A N ,   D O G ,   C H I C K E N   and   *T O R T O I S E*

From links to left:

    - *above*:   Man's embryo = age: 4 weeks, length: 5 mm
                Dog's embryo = age: 4 weeks, length: 5 mm
                Chicken's embryo = age: 4 days, length: 4 mm
                Tortoise's embryo = age: 4 weeks, length: 4 mm

    - *left*:   Man's embryo = age: 8 weeks, length: 14 mm
                Dog's embryo = age: 6 weeks, length: 8.5 mm
                Chicken's embryo = age: 8 days, length: 7 mm
                Tortoise's embryo = age: 6 weeks, length: 7 mm

(after HAECKEL 1868, redrawed)

The human embryo which will conserve the characteristics of a globular brain to its adult form appears as the one amongst the examples shown which has remained closest to the original bipedal type from which they have all developed!

What about a former aquatic stage in man's early evolution?

In 1926, Dr. Max Westenhofer, Professor for Pathological Anatomy at Berlin University made a speech before the Anthropological Congress in Salzburg. In it he declared that apes had originated from the human lineage. Man actually developed from an early water mammal that in turn had developed from ancestors resembling amphibians.

Professor Westenhofer explained that these early predecessors of man were aquatic and that they never passed through a simian stage. The study of the human form should logically, therefore, be carried back to a very early stage in the vertebrate lineage; for instance to the time of the passage from an original full aquatic life to a terrestrial existence.

As I have emphasised in previous articles (see Bibliography), the globular form of the human skull represents the final evolution of an ancient sea-creature's floating organ resembling the umbrella of a jelly-fish.

The Marine Homonculus Hypothesis.

This is an individual development of the the Theory of Initial Bipedalism. Humanity did not develop from a branch of the ape-stock, nor from other quadrupedal antecedents like the Insectivores. According to this theory, the genus *Homo* would have come from a stock peculiar to itself. Thanks to the primitive characteristics of his big, globular encephalon, man has remained a biped.

So, where do we really come from?

Figure two shows the hypothetical reconstruction of the headless marine prevertebrate that might have given birth to the human lineage.

Figure three is a cross section through the body of this same animal that resembled a flat worm swimming in the water. We notice the important features of a dorsal protochorda (sustaining the body structure), the disposition of the muscles and the abdominal (coeliac) cavity in which the absorbed water from the branchial pharynx pours down to the porus. after eventually melting with the genital products that are loosened into the sea for reproduction. Natatory folds on both sides of the body allowed an active and regulated swimming!

The phylogenetical history from the marine pre-hominid forms which THEN evolved up to the four limbed creature with a bony skull is summarised in figure four. We should call this the 'archepagogic' stage.

1. Mouth and Buccal cirri
2. Branchyal pharynx
3. Coeliac cavity
4. Ventral susculature
5. Dorsal susculature
6. Stomach
7. Ailimentary canal
8. Genitals
9. Branchial and genital pore
10. Natatory fold
11. Protochorda
12. Neural duct
13. Anus
14. Bloodvessels
15. Cuticle

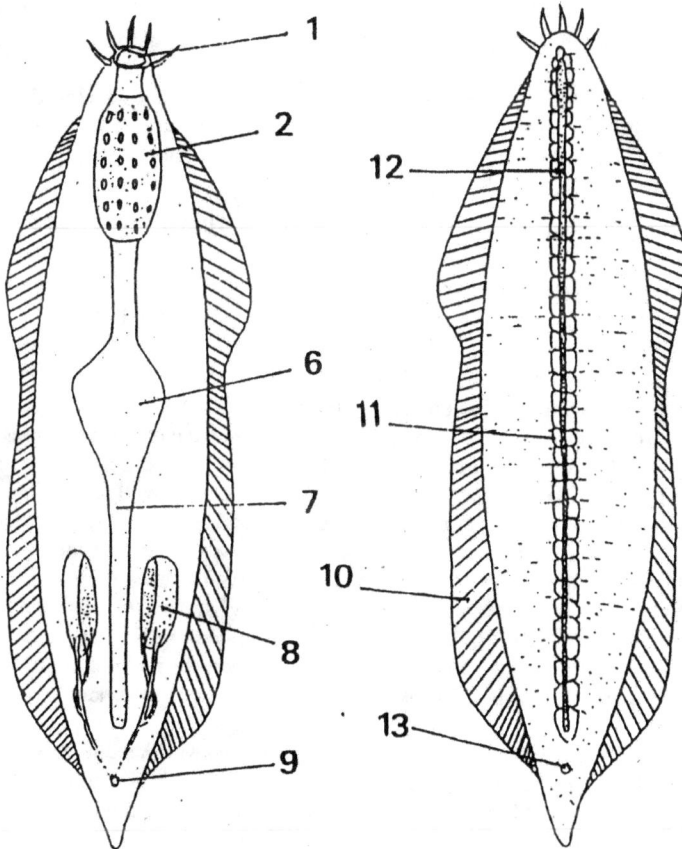

G. 2

HYPOTHETICAL RECONSTITUTION

OF THE MARINE PREVERTEBRATE

Ventral sight (*on the left*) and dorsal sight (*on the right*)

What we may suppose is:

* That a floating organ developed on the apical top of the body of our primitive worm (A) as a 'bubble' like in some medusae. intendedly filled with gas to facilitate an 'up and down' purpose (B).

* That the globular sustenance organ became as functional as, for instance, the natatory bladder of a fish, then compensating for the weakness of the flipper surrounding the body (C).

* That between an inner ectodermal bag filled with gas, and the outer skin, a strengthened mesodermal membrane could then develop. The cells of this would have originated in the protochorda. It was this that shaped the original form and structure of man's skull. (D).

* That such a round configuration as the human skull could only have developed in water. The Marine Homunculus (E) with his four pre-formed limbs and a little tail that functioned as a rudder, then started to evolve into the first ever terrestrial vertebrate.

Pre-Hominids as the first Land-Living Vertebrates.

In remote times this was the completion of the human form, and was also the earliest conception of a *bauplan* for the bipedal-mammals which then represented the earliest terrestrial vertebrate type.

The primordial globular brain which was the direct cause of today's round, human skull. is an organ which developed from the sustenation float of an ancient marine animacule. If we accept this theory then we must also accept that the creature in question also had a functional nervous system that worked before the head itself developed. We will surely find the clues to this ancient nervous system within human anatomy and physiology. These will not be connected with the brain, which at this stage still did not work, but with the spinal chord and the other branchings within the body. This leads us to meridian lines as described in the field of acupuncture. They are indeed tracks of a vanishing system of nervous fibres which represent all that still exists of the former nervous system belonging to our acephalous (headless) aquatic Homonculus. (Figures two and 4a), prior to its evolving into early man. These are the remains of a dorsal nervous system which preceded the brain.

From then on the Homonculus had a brain. This was known as the archepagogic stage, and a representation of this is to be seen in figures 4E and five. This had developed both fore and hind limbs. A higher metabolism developed than that based around the external gills still seen on human embryos.

What was the true appearance of this first vertebrate which conquered the land? We shall soon see.

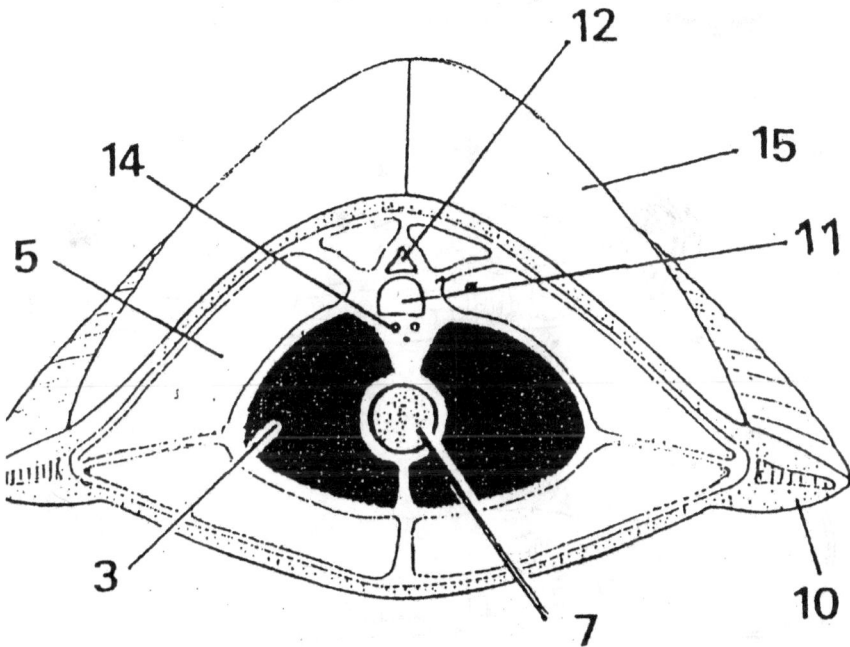

CROSS SECTION

through the middle of the body,

showing the disposition of the musculature,
of the abdominal cavity, the protochorda,
the neural duct and the alimentary canal.

FIG.3

Fig. 4

PHYLOGENETICAL SERIES OF THE AQUATIC PRE-HOMINIDS

Showing the fashioning of the human oesophagus and brain plan through the dwelling up of assustenance organ in a remote marine creature

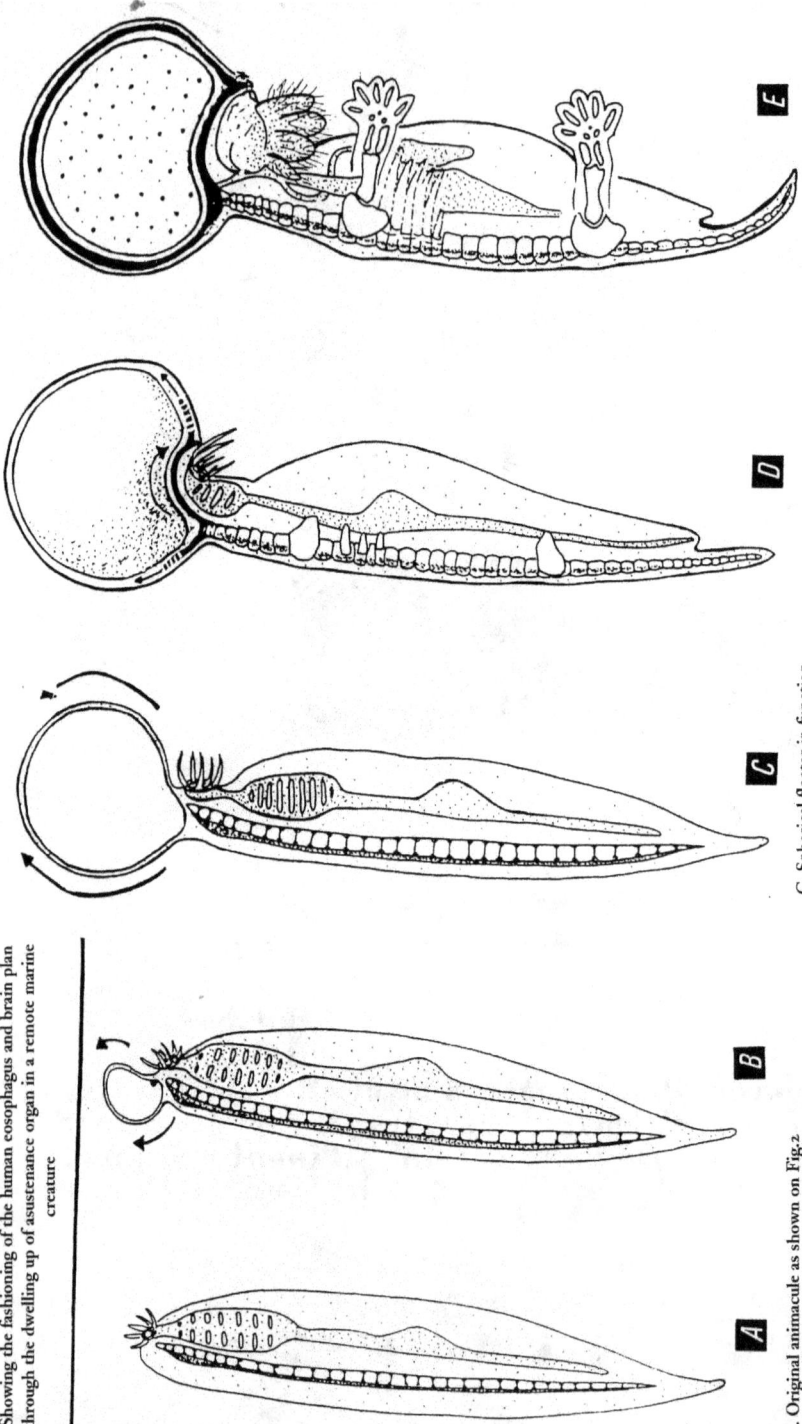

A. Original animacule as shown on Fig.2
B. Beginning of the formation of the apical floater as a sustenance organ

C. Spherical floater in function

D. Function of the partition walls between the 'bubble' - the outer skin; the limbs are developing
E. Primitive globular brain, initial breathing, four webbed limbs and a little tail

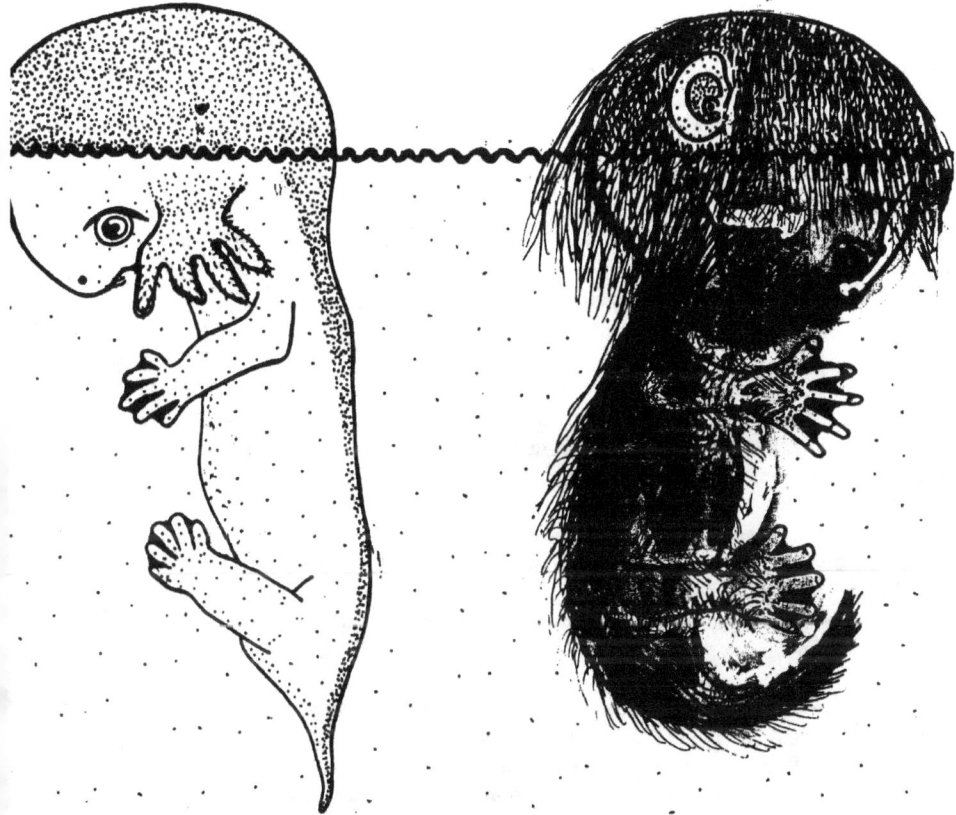

FIG. **5**

ARCHEPAGOGIC - S T A G E S :

M a n ' s   e a r l y

a q u a t i c   e v o l u t i o n

left: original type with branchial resiration, like in
         Fig. 4:D

right: air-breathing (through the ears) haired type,   as   a
         transition to the following amphibious phase

Indeed in this article we shall not try and invoke such annoying problems involved in trying to demonstrate whether a fish that looked like a Coelacanth could really have 'walked' on its paired fins, or whether such a fish might have developed into a four-legged amphibian. because despite the appearance of developing limbs its movement is actually a horizontal undulation like those of a snake.

Let me stress here, that whether the creature was a Homonculus or a coelacanth. there must have been a considerable motivation for an aquatic creature to discover such a new and hostile environment! A small-brained Crossopterygian fish was totally unprepared both physically and psychically to undertake such a glorious adventure.

This truth has been forgotten by most scientists who have, I believe, adopted the gross habit. which is particularly difficult to break, of imagining that at the dawn of terrestrial life something happened, totally different from what happens today when a common tadpole leaves its pool just before becoming a frog.

I have already discussed these conditions, but perhaps it will help to summarise them here. The first land vertebrate had to adapt without any assistance, to a strange new world. Man's insatiable curiosity is indeed an inborn quality as are the inventiveness and exploratory excitement that he still enjoys. We are constantly trying out new ways. and - from the beginning we were an ingenious species.

The first terrestrial pre humans (as depicted in Figure Six) had made decisive steps towards a conquest of the land and made a definitive departure from the ocean where their ancestors had lived. The continuing endeavours of contemporary humans to travel through countries and to reach unexplored places is merely a continuation of this fundamental trend of the earliest pre hominids!

Pioneers of an Undescribable World.

In its archipagogic stage, the marine Homonculus had only developed its brain. formed skull and limbs. The spinal column was also ossified before the animals had left the oceans. Dessication and thermic regulation of the body were therefore the chief problems for a creature that was trying to make huge steps away from the sea!

It was therefore essential to avoid the shrivening of the body. The skin of these first terrestrial vertebrates had therefore to be kept supple, and was therefore covered with an outer horny layer that would restrict the loss of water and with an isolating hairy coat. The sweat glands in the skin intervened in the case of too much heat producing sweat which evaporated to cool the whole body. The sebaceous glands served to lubricate the skin and the hair. All this forms a self-regulatory system able to maintain a stable body temperature in spite of external variations in the atmosphere. Such technic control permitted these marine animals to make a real adaptation to a terrestrial existence.

FIG. 6

EARLY   HUMANS

as they perhaps resembled...

The development of the teeth which allowed the pre-human to chew and digest food quickly is, without doubt, connected with the acquisition of homoeothermy, which needs a great deal of energy. The appearance of hair is also connected with this demand for thermoregulation. Therefore the Homonculus was villous, and this characteristic was also present whilst the creature was still in an aquatic stage.

The once water-dwelling pre-hominid then started to evolve into the first land dwelling vertebrate: truly this was an archaic man.

As I have already emphasised this early creature had an inborn tendency to explore new areas (a pioneer's spirit!), and was the owner of an efficient nervous system. As we have seen, the big brain developed from the marine floating organ. Moreover, our Homonculus already had the pre-adaptations such as bipedal gait, orthograde body position, breathing through the lungs and hairy skin, needed to allow him the loss of his watery habitat as he started to walk on the ground. With his free hands he could pluck and eat plants and some invertebrates which lived on the beaches.

The Homonculus also presented the important characteristic of conceiving and giving birth to living young (viviparity). Before its birth the child grows and develops in the mother's uterus as if it were a portable aquarium! This disposition allowed a complete growth to the delicate big brain and allowed the Homonculus to retain a globular form to its skull, which in turn enabled the creature to retain its bipedalism. Viviparity also allowed the definitive independence of all aquatic contingencies.

The Advent of Humanity!

Bipedality certainly preadapted the early hominids to a life on land. Their normal posture was fully upright, and the head was resting without muscular strain on the shoulders. Derived aptitudes are those of the non-human vertebrates, as they modify the carriage of their head and the posture of their body (dehominization).

Why MUST two legged creatures with big heads have been the first mammals that ever lived? Simply, because of the embryological facts (see figure one again), as the scienific research indicates. It is worth noting that the globular form of the early encephalon controls the flexion forward of the anterior part of the embryonic spinal chord and so maintains our upright posture. The globular form of the skull is the most primitive in embryos. This feature is consequently the most ancient in terms of phylogenetic rapprochement, at least as far as the actual history of the vertebrates is concerned.

So, in reality, modern man evolved from .... an early man! This was without an intermediary, and they would have differed slightly from the older forms as described above. They might even have been a little taller!

The Palaentological discoveries of 'our ancestors' concern either ape-like creatures (*Australopithecus, 'Homo' habilis*), or deal with specialised post-human lineages such as *'Homo' erectus* and Neanderthals. True man existed prior to them. He was a contemporary of the armoured fishes and the huge quadrupeds and he even fought against the dinosaurs.

I am well aware that there are some authorities who will fervently disagree with what this article states about an early aquatic origin of humanity. Above all, I want to assert that a free thinking spirit is necessary in science, particularly as regards our provenance as a species, despite the emotional trends which are evidently at work!

BIBLIOGRAPHY.

BONIK, K. (1976): *Die Evolution der Tetrapoden als Problemage*. Natur und Museum. 108 (5): 133-136, Frankfurt.

FRECHKOP, S. (1941): *Remarques sur l'embryologie des Mammiféres*. Bull. Mus. Roy. Hist. Nat. Belg., 17: 1-6.

HARDY, A. (1960): *Was Man more aquatic in the past?* The New Scientist, 7: 642-645, (London).

HEUVELMANS, B. (1954): *L'Homme doit-il etre considre comme le moins specialise des mammifres?* Sciences & Avenir, 85 (3): 132-136, Paris.

de SARRE, F. (1989 a): *About an aquatic stage in Man's evolution, with references to Max Westenhofer and Initial Bipedalism*. Bipedia, 2: 1-4, Nice.

de SARRE, F. (1989 b): *La theorie de la Bipede Initiale (2nd Part)* - 3' Millenaire, 14: 35-50, Paris.

de SARRE, F. (1992 a): *La Coaelacanthe a-t-il un passe terrestre?* Bipedia, 8: 1-5. Nice.

de SARRE, F. (1992 b): *kamen unsere Vorfahren aus dem Ozean? Ueber die aquatile Lebensweise des Menschen in den fruhen Zeiten seiner Entwicklung*. Efodon News, 11: 13-15, Russelsheim (Germ.)

de SARRE, F. (1992 c): *The Marine Homonculus Hypothesis, an alternativve paradigm for human earliest evolution* - Bipedia, 9: 13-16, Nice.

WESTENHOFER, M. (1935): *Das Problem der Menschwerdung*. Medizin. Welt, 31 to 44, Berlin.

*****

Mysterious Monkeys of Hong Kong

by Jonathan Downes and Richard Muirhead

AUTHOR'S NOTE: We have been working on our book about the mystery animals of Hong Kong for about three years now. The project is taking far longer than we had originally thought and seems to be reaching epic proportions. However, it will eventually be complete. In the meantime here are some, almost random, jottings of the simian kind!

The first connection that would be made in the minds of most people between monkeys and Hong Kong, is the widespread belief in the Chinese custom of eating raw monkey brains, often out of the trepanned skull of a living monkey. It is interesting to note that whereas some television documentaries shown on British TV during the late 1980's repeated this story with glee, others denied that this was more than a historical curiosity that certainly doesn't happen any more.

The present authors believe that whatever perversity can be dreamed up by the human psyche is catered for somewhere on the globe, and as most human perversities are catered for somewhere in Hong Kong, monkey brains are probably still scooped out and eaten somewhere in the less well traversed areas of the territory.

The living monkey population of Hong Kong is no less mysterious. Although there is no doubt that monkeys have existed in Hong Kong since at least 1918, and probably have always existed here their status is far less certain.

It is interesting, given the speculative nature of most of the simian fauna of Hong Kong, that Brown and Lee, writing as recently as 1993 refer to 'the mythical monkeys that are said to frequent the trees' of The Peak, in an otherwise matter of fact tourist guide to the territory. (P. 79)

Even the experiences of the two authors differ greatly differ highly. Jonathan Downes lived in Hong Kong between 1961 and 1971, and only saw a monkey once, in 1970 or 1971 when it was crossing the transverse cables of the Peak Tram somewhere above Barker Road.

Richard Muirhead, however lived in Hong Kong between 1966 and 1985 and saw monkeys on a number of occasions in a number of different localities. He saw them at Mount Kellet Road leading down to La Hacienda on the Peak.

Hong Kong and Vicinity

The lists of mammals recorded from Hong Kong varies wildly according to the source and the date. Each of the references cited gives a different species or combination of species as resident in the territory, with reports only ten or fifteen years apart varying dramatically.

Perhaps we should now examine the records of monkey species in more depth:

In 1870 Swinhoe described the local monkey as *Macacus Sancti-johannis* and wrote:

"This rock monkey is found in most of the small islands about Hong Kong and is like a Rhesus with a very short tail".

He continued:

"Dried bodies of this animal split in two are often exhibited from the ceiling in druggists shops in Canton and Hong Kong; and its bones are used for medicinal purposes".

Writing in 1951 Herklots reported:

"There are still monkeys wild on the Lema Islands south of the Colony. On the island of Hong Kong a monkey family was watched early one morning near Tai Tam reservoir in 1947, and I have had occasional accounts of monkeys having been seen on The Peak and in the Deep Water Bay valley. It is possible, but not certain that these are descendants of the original wild stock".

Herklots identified the original wild monkeys as Rhesus Monkeys, which, at least when he was writing had a range stretching from *"India to the whole of China south of the Yangtze"*.

This is completely at odds with the facts, even those that he presented himself. Swinhoe had recorded in 1866 that monkeys were present on:

"Most of the small islands in the bay of Hong Kong",

but his description of *"M. Sancti-johannis"* specifically said that the Hong Kong monkeys were like a rhesus but with a short tail. The rhesus monkey has a relatively long tail, and was first described by Zimmerman in 1780. It is an animal with which Swinhoe was bound to have been familiar, and therefore if he drew a distinction between the 'Hong Kong Rock Monkey' and the rhesus then we can safely assume that they were two different animals.

Herklots continues:

"The monkeys that live in the woodland near the Kowloon reservoir are not descendants of the wild stock but of monkeys released during the first world war. During the Japanese occupation of the Colony during the second world war, after the trees had been cut down, the surviving animals scattered. Since the war they have been reported from several districts in

Rhesus Macacque
(Macaca mulatta)

the New Territories including their old haunts".

Herklots appears unsure. and implies that these animals from the Kowloon reservoir area are not related to the animals reported from Hong Kong Island itself. He was, however optimistic about the recovery of the species.

"With the growth of the trees planted by the Forestry Department in the reservoir area they are likely to return and stay, especially so if fed by the public as their parents used to be".

Herklots' optimism, was, we shall see, to be justified. His feelings about the relationship between the Hong Kong and Kowloon monkeys were corroborated by the next report by Patricia Marshall, only sixteen years later. Marshall, reiterated what Herklots had to say but insisted that the animals were of a different species again, the Long Tailed Macaque (*Macaca irus*). She wrote:

"The group of monkeys living in the Kowloon reservoirs catchment area are long-tailed macaques which were probably released or escaped from captivity during or shortly after the 1939-1945 War. The rhesus monkey (Macaca mulatta) which was once abundant on Hong Kong Island and in the New Territories now appears to have died out".

This is in complete contrast to Herklots who says that the same troupe of monkeys was liberated around about the time of the first world war. Marshall showed that Herklots was justified when he said that people would feed these monkeys, however and suggested that it was better for people not to feed them, because otherwise they soon become tame, and some people can misinterpret their actions. This has...

"...led to a number of monkeys being shot including some very tame ones on Hong Kong Island".

This suggests that the 'demise' of the island population may not have been that far previous to the time when Marshall was writing. Only three years after Marshall wrote that the island population was extinct Jonathan Downes saw a monkey on the Peak Tram, and within a few years Richard Muirhead also saw these animals on the island itself.

In 1981, fourteen years after Marshall's book Hill and Phillips included both species in their list of Hong Kong Mammals. They wrote:

"On Hong Kong Island the usual monkey to be seen in the area of the Peak, Pok Fu Lam, Stanley and Tai Tam Reservoir is the Rhesus Macaque. This species, frequently used for medical research. is characterised by having a shorter tail, (about half the body length), a pink face and in the male, conspicuous red buttocks. They have been seen in a group at Tai Tam, but elsewhere on Hong Kong Island single animals have mostly been observed. A small group is now established in the Tai Po Kau Reserve, New Territories. and another one or more groups around Kowloon Reservoir".

A9MD42 Alamy Images

Long tailed macacque

This claims not only that Herklots' original theory was right but that the Hong Kong Island animals had become a thriving population from nowhere in only fourteen years. Obviously Herklots, Marshall and Hill/Phillips could not all be right.

Hill and Marshall continue:

"Two species of Old World Monkeys are now established locally although their previous occurrence as part of the indigenous fauna is not at all clear. At present populations here are thought to have come almost entirely from escaped or released animals which eventually have formed small breeding colonies.

At least twenty Long-tailed Macaques are present in the wooded area around Kowloon Reservoir. They are characterised by having a tail about as long as the body. In other parts of S.E.Asia this species can commonly be seen on seashores where they catch crabs and molluscs. Sometimes called the Crab Eating Macaque it also eats wild figs, various fruits, leaves, insects and picnic remains. Adult males are large and aggressive, and weigh 6-7 kg. The face is generally quite furry, and they do not have the naked red buttocks".

We now have two different species of monkey in the Kowloon Reservoir area. The animal seen by Jonathan Downes in 1970 had the red buttocks proving that it was a Rhesus monkey. Hill and Phillips imply that Rhesus Monkeys are the only ones to live on the island itself, but even they are unsure of the provenance of the Hong Kong monkeys.

We still need to consider Swinhoe's original record. *Macacus Sancti-johannis* is said to have a very short tail. As we have seen both the species that were found in Hong Kong before 1981 have quite noticeable tails.

If we are to accept Herklots' assurance that the Kowloon animals were liberated during the first world war, we are still left with the anomalous animals of the island itself. These do appear to be Rhesus monkeys, but where were they between 1947 and 1970? We are also left with the riddle of *Macacus Sancti-johannis*. At least one authority claims that this is an animal species, like the Hong Kong newt, unique to the territory.

Writing in 1893, Sydney B.J.Sketchley claimed that:

"Our island shows no trace of having been submerged since remote (say Silurian) geological times, and though its severance from China may have taken place as late as the Tertiary period, it is as an island newer than Formosa or Hainan.

When it was insulated we cannot exactly say, but if we judge by analogy from the adjacent Ladrone islands it was long enough ago to admit of the evolution of a distinct pig tailed rock monkey (Macacus sancti-johannis), whose dried bodies may be seen in Chinese medicine shops".

This description of a 'pig tail', fits neither the rhesus or the Long Tailed Macaque, although it does fit several other S.E.Asian species such as the Stump Tailed Macaque *(M.arctoides)* and the Tibetan Macaque *(M. thibetana)*.

Goodyer's 1992 mammal list confused the matter further. He reiterated many of the references already cited, but in the interests of completeness, we quote him in full:

"There are at least four species of Macaques in the territory. The Rhesus Macaque Macaca Mulatta (Zimmerman 1780), the Crab Eating Macaque M.fascicularis (Raffles 1821), and small numbers of the Japanese Macaque M.Fuscata (Blyth 1875) and the Tibetan Macaque M.thibetana (Milne-Edwards, 1870). Interbreeding occurs between M.mulatta, M.fascicularis and M.fuscata but not with M.thibetana.

Herklots (1951) stated that the macaques living around the Kowloon Reservoir were released during the first world war, but goes on to say that they used to be common on most of the small islands in the bay of Hong Kong in the late nineteenth century. Apparently the macaques in Hong Kong today derive from reintroduced stock and there are about seven hundred in the Territory. (J.Fellowes, pers. comm). The majority are confined to the Kowloon Hills in the Kam San and Lion Rock Country Parks (see Romer 1966; Southwick and Southwick 1983). There are also groups at Shing Mun and Tai Po Kau. Repports of indioviduals, usually young males, have been received from various parts of Hong Kong.

The Macaque population of the Territory is growing rapidly and a study of the associated problems has been carried out by the World Wide Fund for Nature (Fellowes 1991). Also in this Memoir Fellowes comments on the natural diet of Hong Kong Macaques".

A 1988 television documentary about the animals of Hong Kong, also included a section about the Macaques:

"Relatively few people in Hong Kong venture outside the cities. The idea of mountain crabs and monkeys being found living just beyond the outskirts seems almost as unlikely as the idea of finding live dragons. Macaques were probably native here but were wiped out many years ago by the Chinese. The present populations are the descendents of subsequently released animals which established themselves in the wild.

Like most monkeys macaques are inquisitive and intelligent creatures. They quickly become 'streetwise' and it is not long before thgey lose their fear of human beings. The local people enjoy feeding them even though adult males have a reputation for biting. In recent years they have become a nuisance at picnic spots, they readily accept free handouts and soon discover the best sources of food.

Even Crab-eating Macaques rarely bother with more natural food these days... "

The film segment which lasts something in the region of two minutes shows what appears to be a large family group of monkeys playing and scavenging around a picnic site somewhere in the New Territories. There are signs around which read, in Cantonese and English:

"WARNING. Do not feed or agitate the monkeys. They may attack you and spread infectious diseases".

What is particularly interesting about the animals in this particular documentary is that they exhibit a wide range of characteristics. Unbfortunately for the hapless cryptozoologist hoping to find a 'pig tailed' macaque amongst them they all have tails, but some are the length of the body, signifying that they are, indeed Crab eating Macaques, as identified by the commentary, and some have shorter tails, half the length of the body, identifying them as Rhesus Macaques. Some animals have vaguely reddish buttocks, but the full red buttocks of an adult male Rhesus monkey are not to be seen. It appears likely that these are either the hybrids identified by Goodyer, or, more interestingly, proof that the two species not only interbreed but interact on a social level as well.

Acccording to the noted zoologist C.H.Keeling, *Macacus Sancti-johannis* is generally considered as a sub species of the Rhesus Monkey. The type specimen, a female, was presented to London Zoo on the 14th January 1867 by R. Swinhoe. It had been caught on North Lema Island, just south of Hong Kong territorial waters. It is said to be found on St John's island in the South China Sea and both the monkey and the island were named after Commander St. John R.N.

It is dissappointing to discover that *Macacus Sancti-johannis* is not distinct at a specific level, but the mystery still remains. If this subspecies is different enough from the Rhesus Monkey to be accorded sub-specific status, and if the tail is so short that Sketchley claimed it was 'pig tailed'. then what actually happenned to the Hong Kong populations?

The animal must have occured on Hong Kong Island in 1893, because Sketchley implied that it was living in British Territory. At that time only the island itself and a small portion of what is now Kowloon City were under British control.

The title of Sketchley's book, also implies that the animals were found on the island itself. It seems likely that these were the ancestors of the animals described by Herklots only fifty four years later.

Even if we ignore the mystery of where the Hong Kong island animals disappeared to between 1947 and 1970, and even if we ignore the fact that Goodyer's records and indeed those of John Fellowes, writing in his WWF Briefing Document, (1992), do not mention monkeys on the island itself, there seems no doubt that Rhesus monkeys are, and have been resident on Hong Kong island.

Whether or not we are to accept that Macacus Sancti-johannis is distinct merely at sub-specific level, and the interbreeding between three of the four species described by Goodyer in

1992 suggests that there is more to be learned about the genetics of the Macaque family, then the animals were certainly distinct enough morphologically with a 'pig-tail', (suggesting a short, curly tail), that they were for a while considered to be a distinct species.

It seems likely that the genes which produce this 'pig tail' are hidden somewhere within the gene pool of the Rhesus monmkeys on Hong Kong Island itself. It also seems most probable that these genetic differences may yet assert themselves once more and that the zoologists of the future should be prepared to echo the words of Sketchley in 1893, and assert that Hong Kong Island is once again the home to a distinctive pig tailed monkey!

REFERENCES

BROWN, J & LEE, H. 'Hong Kong and Macao - the rough guide' (Rough Guides, London 1993)
WELSH, Frank 'A History of Hong Kong' (London, Harper Collins, 1993)
SKETCHLEY, S.B.J. 'Our Island - a naturalists description of Hong Kong' (Hong Kong. Kelly and Walsh, 1893).
HERKLOTS Dr. G.A.K. The Hong Kong Countryside. SCMP 1951
MARSHALL, P. Hong Kong Mammals. H.K.Government Pubs 1967
HILL D and PHILLIPS K. The Animals of Hong Kong in Colour. H.K.Govt. Pubs 1981.
GOODYER, N. The Mammals of Hong Kong. H.K. Nat. Hist. Soc. 1992.
FELLOWES, J. Hong Kong Macaques. WWF HK 1991.

* * * * *

A Record of Rare and odd-coloured animals in Norfolk.

by Justin Boote.

(EDITOR'S NOTE: The following article was so impressive that
we have decided to commission similar papers covering the
rest of the United Kingdom. Should you be interested in
preparing a paper on the area in which you live, please
telephone me at the Editorial number. Eventually we intend
to gather all such research together in a book).

The following information is based upon the Norfolk Bird and Mammal Report which is issued every year and has done so since 1953. The mammal report began in 1959 and still continues with a few years missed out due to financial restrictions.

When the reports began there were 61 species of mammal on the Norfolk list. There are now about forty, and many of these are now, sadly, uncommon. The county's most famous animals, the red squirrel and the otter are very rare, and indeed the otter may now be extinct in Norfolk. or at least the vast majority of it.

Although many of the county's animals are rare, Norfolk can still be proud of its deer species which include almost all those found in Britain. Many are to be found in the Thetford area which also boasts the largest area of pine forestry in the U.K. This forest provides a safe refuge from Man and their other predators for many of Norfolk's known. and sometimes unknown animals.

MAMMAL REPORT FOR 1959

The problem with compiling a mammal report is that many of the more common animals are overlooked or ignored, deemed unimportant. Accurate and detailed research is therefore almost impossible. The hedgehog is a classic example of this, yet colour variations are rare in the county. Only one was recorded during 1959 - an albino at Bodham. No date was given.

The mole is another animal often overlooked, although colour variations are more common. Three were seen during 1959 and these were white or cream coloured. These were seen at Fakenham. Barnham and Old Costessey.

Like many animals the brown rat has an unenviable and unpreposessing name, yet this too offers some interesting colour variants. Four albinos were seen at Bressingham and one at Roydon. There were also sandy coloured rats at Thornage, letheringset and Lynford.

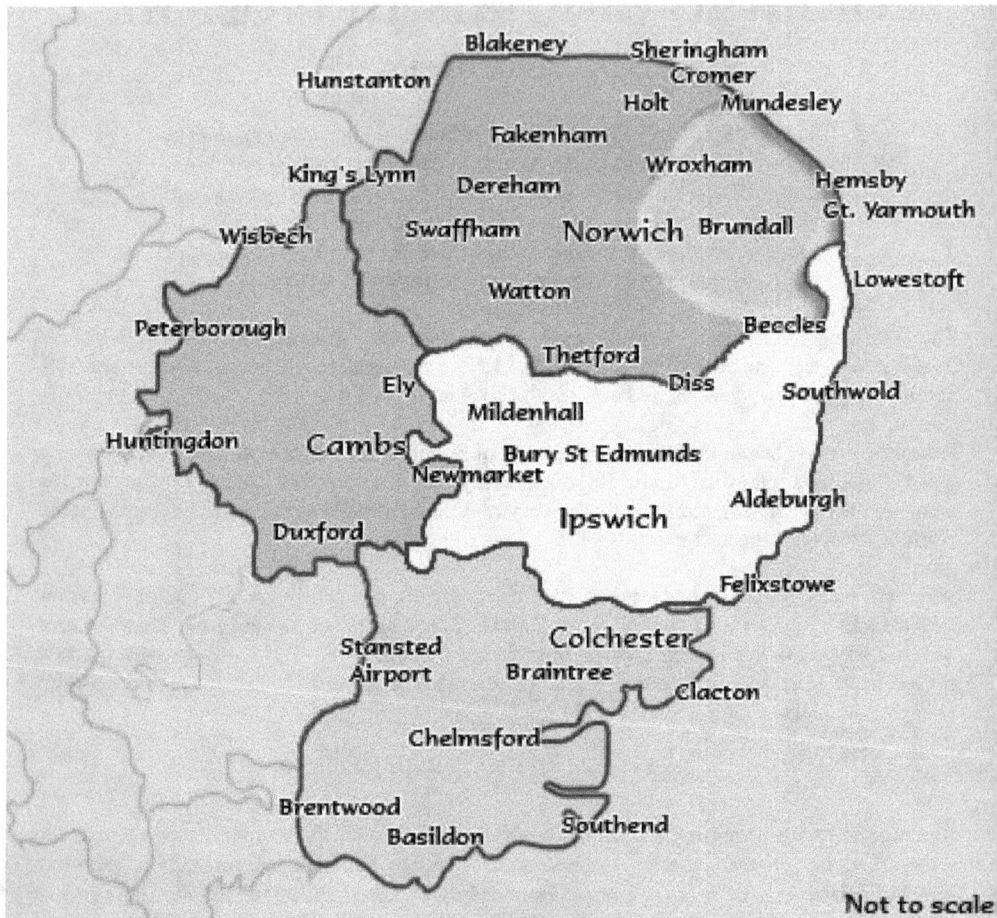

Blakeney · Sheringham
Hunstanton · Cromer
Holt · Mundesley
Fakenham
King's Lynn · Wroxham · Hemsby
Dereham · Gt. Yarmouth
Wisbech · Swaffham · Norwich · Brundall
Watton · Lowestoft
Peterborough · Beccles
Thetford
Ely · Diss · Southwold
Mildenhall
Huntingdon · Cambs · Bury St Edmunds
Newmarket
Duxford · Ipswich · Aldeburgh
Felixstowe
Stansted · Colchester
Airport · Braintree
Clacton
Chelmsford
Brentwood
Basildon · Southend

Not to scale

Red squirrels were still quite common in the county at this time and can be found at several locations. Several years before. a small colony of melanistic animals of this species lived at Ellingham but by 1959 they had almost died out. During the year only one melanistic animal was seen at Ellingham Hall.

Norfolk was home to a family of feral domestic cats during 1959. Five were known at North Walsham woods. one with several kittens was seen at Thetford Chase and others are known at Horsford.

There were at least ten mink farms in the county, and it is therefore not surpreising that several have escaped. One captured in Yarmouth. had travelled about ten miles from Somerlayton. In August another was shot at Thomson Mere. which is many miles from the nearest mink farm. This demonstrates the long distances the mink are prepared to travel in order to find a suitable habitat. This particular animal was a male and about three years old. It measured two feet in length with a six inch tail. Another was seen twice at Lynford Forest and there was a reported shooting of a mink at Winterton.

Cetaceans are quite rare along the Norfolk coast and lesser rorqual whales are even rarer. The last time one had been recorded prior to 1959 was on the 13th October 1943. One was found during 1959 stranded on the beach between Waxham and Sea Palling. It measured sixteen feet and seven inches in length with a body circumpherence of nine feet, a tail span of four feet and it weighed two and a half tonnes.

MAMMAL REPORT FOR 1960

There were not too many reports for 1960. although this was somewhat tempered by the return of an animal unrecorded for nearly forty five years.

There were no records of oddly coloured hedgehogs this year, but quite a few aberrant moles were noted. Twelve moles with yellow/orange fur on their bellies were observed at Kelling along with two completely yellow and orange. Another golden mole was recorded at Merton.

. Norfolk has a large population of rabbits around the county and several of these are of a mixed colour. mostly white and black. They seem to be most common in the north. An albino was seen at North Walsham and Shotesham. and black rabbits were seen at East Carlton and Binham. It is unknown if they were truly wild animals or whether they originated from escapees.

Water voles are also common in Norfolk and, again there seems to be a higher proportion of melanistic voles in the north of the county. Three were seen at High Kelling and there were others at Corpusty, Bawdeswell and Hethersett. They are apparently very common in the Broadland marshes and resemble very closely the sub-species A.a.reta which is found in Scotland north of the Clyde and Tay. It is known. however. that the skull of the black form found in Norfolk more nearly resembles the skull of *A.a.amphibius*, and it seems unlikely that the Norfolk water voles are a surviving population of *A.a.reta*. The dominant black mutation appears to be an ancient one but its survival significance is unknown.

Red deer are common in Norfolk and many fine specimens can be found all over the county. Albinos are not so common in Norfolk or anywhere. However, one was seen at Castle Rising near King's Lynn in July 1960. It was with two other bucks which were of the normal colouration.

By 1960 the muntjac had been slowly edging its way towards Norfolk for some time. The first record was in 1951, from Walsingham. Others followed from Lowestoft in 1952 and Thetford Chase in 1953, and there were unconfirmed reports in 1958 and 1959. In 1960 an undoubted muntjac was shot at Holkham Hall and more were expected.

(EDITOR'S NOTE: It is unclear whether this was *Muntiacus reevesii* or *M.muntiacus*. Both species now live in Britain.)

There were a few cetaceans in 1960. Indeed there were slightly more reports than there had been in the previous year. Four strandings of porpoises were recorded at Colt Head, one in July and three in June. Another was stranded at Cley in October. A solitary dolphin (species unrecorded) was found dead on the shore at Waxham on Christmas Eve.

There were a few reports of animal attacks during the year:

At Langley, an otter was alleged to have attacked a man, and later one was caught weighing 26lb and measuring 48 inches in length.

At Corpusty in March, a violent fight took place between a weasel and a stoat. It lasted over ten minutes and the participants were frequently locked together and rolled over and over in a mass like a furry ball. There was a great deal of hissing and squeaking. The stoat eventually carried off the limp body of the weasel.

Norfolk also recorded its first lesser horseshoe bats since they had been recorded at Happisburgh in 1914. Two were captured in the east. The exact location, however, is not given!

MAMMAL REPORT FOR 1961

1961 was a good year for colour variants amongst the Norfolk mammalia. Again, no aberrantly coloured hedgehogs were seen, which demonstrates how rare they are, but twospecimens from Kelling and Weybourne were found badly affected by a form of Ringworm. Both had completely bare patches of skin on their backs, each about one and a half inches in diameter.

Moles with unusual colouration patterns were again seen in good numbers. An orange coloured variant was seen at Thuriton. a cream mole from Costessey and a brownish-grey one from Hethersett. The white form was also seen. This came from Bressingam. in the south. This is the first record of a white mole from the southern part of Norfolk.

The mammal with the most common colour variant seems to be the water vole. It was estimated in 1961 that in the north of. the county almost 50% of the population were melanistic. Black specimens are also common in Aylsham, in the lake at Bolwick Hall. Marsham and in the Yare. Very dark brown specimens were recorded from Brooke. Wroxham. Trunch and Hathersett.

An interesting peice of behaviour was observed at Blickling during July. A water vole. while under observation. was seen to remove a water lily and swim with it, by carrying it above its head. apparently as a form of camouflage.

The rare dark form of red squirrel still appeared to be surviving in Ellingham woods. despite the pessimism expressed two years earlier, as another was recorded at Ellingham Hall. There appears to be no evidence suggesting that these animals were introduced specimens of the darker coloured central European race, and as far as is known these black-looking specimens are natural melanistic mutants.

A dramatic decline. in the stoat population during the year was also noted, with many being shot by farmers as pests. A few with black and white striped faces were seen at Taverham. one brown and white one was observed at Glandford in late December, and one in Ermine was noted in the Breydon marshes.

In contrast to the above species, it appeared that the weasel was on the increase in Norfolk and many road kills were noted during 1961.

A completely white, but not albino weasel was seen on several occasions at Swaffham Forest. and a very light ginger specimen was taken at Shipdham in April. Another, not quite so light. was seen at Brisley in May and a weasel with a white forehead was seen at Brinton.

(EDITOR'S NOTE: We have published a great deal about aberrant colour variations in mustelids. most notably in *Animals & Men* issue One and in parts of my book *'The Smaller Mystery Carnivores of the Westcountry'*. In that book I discussed the reports that ferrets had been succesfully interbred with stoats. The reports of stoats with 'striped' faces, could be seen as corrobvorative evidence for such unusual matings in the wild!)

The mink was still on the increase during 1961, and the *Aylsham Rabbit Clearance Society* caught two. Another entered a coypu trap at Oxnead.

The numbers of feral domestic cats in the county were also on the increase. At Witton and

Thetford they were reported to live in burrows and rabbit holes. They were also said to inhabit thickets and woods at Aylsham, North Walsham, East Bilney and Bylaugh, as well as at the Horsford-Swardeston area, living off small mammals and rabbits.

A large school of porpoises was seen on a number of occasions. On June 23rd it was so big that it took ten or fifteen minutes to pass. It was going east, off the coast at Cromer, and they were only a few yards from the shoreline. There were several strandings as well, with incidents at Sheringham in July, Cley in September and two more at Holkham.

On October 3rd at least six were observed at Sheringham within 100 feet of the shore and another was found stranded at Yarmouth's North Beach in December.

There was also a solitary report of the white-beaked dolphin, a juvenile, six feet long at East Runton on April 29th.

MAMMAL REPORT FOR 1962

A new variant was recorded this year with the discovery of a dead, melanistic common shrew at Costessey, and even more unusually; a shrew was seen at Taverham which had a greyish head, neck and hind quarters and with a dark middle down its back.

Odd-coloured moles were again seen on several occasions. A mottled orange coloured variety was seen at Lakenham Old Hall in February, a creamy golden one at Docking in early May, a pale cream one at Salthouse in September and a light buff coloured one at Swanton Novers in mid-December.

One mole was seen swimming strongly for fifty yards in Bayfield Lake.

There were not as many records of variant rabbits during 1962 as might have been expected. The only records are of black animals seen at Westing Heath and a solitary black animal at Morton.

Melanistic water voles were still thriving in northern Norfolk during 1962, where the ratio was said to be one in five of the population in the Glaven valley. A solitary animal was seen twice in the moat at Caister Castle.

The black rat was almost extinct in Norfolk in 1962 but the numbers of brown rats had increased dramatically over the previous year and a half, and it was noted that the population had almost reached epidemic proportions. One nest in Happisburgh contained a litter of sixteen young rats.

Only two colour variants were recorded from this species, however, and they were both albinos. One was a male at Branthill Farm, Holkham on November 10th, and one, a white female with purpleish eyes was killed at Boston Hall exactly a month later.

It was noted that during 1962 the red squirrel population which had been very healthy for some years, appeared to be on the decline in many areas.

In north Norfolk it was the worst, with many found dead or dying for no apparent reason in Swaffham forest. In Horsford, a large company of about twenty squirrels were observed all active in one large oak tree, the significance of which is unknown.

The decline of the stoat in Norfolk appeared to be slowing down during 1962 with several areas actually noting an increase. Only three variants were recorded, and these were all in ermine. One was shot at Fakenham in late February, and another was seen in the same month on the river in the same location. A partially white specimen was recorded at Weston Longville in April.

Although it is generally believed that weasels do not turn white in winter, several instances of white, or partially white weasels have been noted. A weasel with dark eyes and a completely white pelage was noted at Valley House, Holt in February.

There were no records for mink during 1962, but feral domestic cats were noted from several new locations. They were recorded from Whitwell, Wramplingham, Bilney and Taverham. It was noted that their numbers had increased dramatically at Horsford.

Several species of cetaceans were reported during 1962. Three stranded porpoises were discovered at Holme in March, a female at Cromer on April 21st, and a ten foot specimen at Sea Palling. Several were seen close in shore at Burnham Ovary in late August.

A fully grown white beaked dolphin was stranded at Holme, and another, about seven feet long, passed within a few yards of bathers at Holkham in early September.

Two lesser rorquals were observed at King's Lynn in February and at North Wooton Marsh in the same month. Two bottle-nosed whales, each about seventeen feet long were washed ashore on July 10th at Holkham. This species had not been reported since 1951.

MAMMAL REPORT FOR 1963

There was a huge increase in moles during the year, all over the country, and there was the usual collection of colour variations as well. A sandy-cream variant was taken at Stanhoe, an orange coloured one at Costessey in late May, and several moles with orange markings on the chest and abdomen were caught in the Bawdeswell area. One very large mole measured 15.5 cm from nose-tip to tail-tip. A light buff-coloured one was seen at Swanton Novers churchyard and a silver-grey variety was taken at Bodham woods.

The rabbit was another species that made significant increases in population during the year despite the threat of myxamotosis. In the Breckland area two white animals were seen which is somewhat of a rarity in that area. Another white one was seen at Binham. A black specimen was also observed at Holme bird observatory.

The first variants of the hare within Norfolk, were reported in 1963 with a white animal caught by a dog at Middle Drove, Marshland St. James in early October. A second aberrant hare - an unusual grey coloured specimen was shot at Downbland Farm, Wymondham.

There were more reports of water voles. These were seen at Bawdeswell, Gressenhall and Kelling. The proportion of black to brown is thought to have decreased in North Norfolk.

There was just one variant rat during 1963 - an albino at North Creake. There was some consternation concerning the red squirrel following the discovery that many of the young are infested with fleas. One drey containing three recently born young was so infested that no part of their bodies was free of bites. A young squirrel disinfested at Gunton was found to have over 150 fleas on its body.

It was noted that the grey squirrel was moving closer to Norfolk, and concern was stated that they may breed in the county if able. They came mostly from the south and south-west as pairs were seen at Tuddenham Heath in early September, another one at Chippenham Fen in the same month, and a third pair at Elveden in early October. Others were seen in the west of the county. It was noted that this was a serious problem for the Forestry Commission as they can influence the population of the native red squirrel. They are also regarded as major pests.

Due to the hard winter of the previous year there were more reports of stoats in ermine. One was seen at Attlebridge in January and others in the same month at Thetford and Aylsham. In February, one was shot at Wramplingham, another at Swannington and another hunting rats at Haddiscoe. An almost completely white one, save a brown patch on the head and mid dorsal line was discovered at Erpington in February, along with two others in full ermine on the 27th. There were other reports from Watton and Bungay.

Later in the year a stoat was observed 'dancing' inside a ring of blackbirds, thrushes and redwings.

The mink also made it through the winter without too much hardship, indeed there were more reports during 1963 than there had been in previous years. Six were taken in the Thetford-Brandon area and another three in the Shadwell-Brattenham area. Some were still known to be at large at the end of the year. One species that didn't survive the hard winter were the feral domestic cats which were, apparently, wiped out in 1963.

There were very few reports of stranded porpoises during the year. One was recorded at Winterton on January 10th. It measured four feet and eleven inches. Remains were found at West Rutton in August and at Holme in January. The hind part of a white-beaked dolphin was washed ashore during January gales at Runton.

MAMMAL REPORT FOR 1964

There was just one record of an oddly coloured mole this year - an orange coloured animal. No date was given. Old-time molecatchers are reputed occasionally to have caught enough albino moles to have made themselves a white mole-skin waistcoat.

There were no reports of albino rabbits in 1964, but a black specimen was shot at Binham in November. These melanistic forms must have been at one time fairly common in Breckland, where there is at least one heath known as 'Black Rabbit Warren'.

Unusual colour variants in hares are quite rare considering the population of these species in Norfolk, but there was an interesting sighting on the Norfolk/Suffolk border during 1964. It was *"basically white, with a large black mark on the back, slightly left of its mid-dorsal line"*.

The melanistic water voles were as common as ever and there was a new location that year with the discovery of two dead animals at Wymondham College in July, some distance from water. There was also the usual population at Corpusty.

The brown rat population dropped slightly during the year although they were still very common. There was only one variant recorded during 1964 - a piebald female. There were no reports of the black rat during the year, a fact for which the mammal recorder was 'thankful'.

A fox shot near Yarmouth was found to have the left hind leg one and a half inches shorter than the other, and half the pad was mising. The bone had broken, impacted and reset itself. A 21lb dog fox, four feet and two inches long was run over in Horsford in October. The stomach contents of one found dead near Thornage was found to contain several species of fungi.

Stoats in ermine were one of the zoological highlights of 1964 in Norfolk. Several perfect, or near perfect ones were seen in North Norfolk, where they are said to be increasing generally. Ermine specimens were also seen at Attlebridge, Aylsham and Thetford Forest in January, and at Wramplingham, Swannington and Haddiscoe in February.

There was just one report of mink during the year, when a specimen was observed at Santon Downham.

The one record of wild cats comes from Horsford, where they are stated to be continuing to increase, living in trees and old rabbit burrows. There may well be others - there are certainly many cats about whose lives are only tenuously connected with human habitation.

(EDITOR'S NOTE: The mammal Recorder almost certainly meant feral cats rather than the indigenous British Wildcat which was hunted to extinction in Norfolk in historic times.

Whilst I have gone on record as saying that I do not accept all the findings of Messrs. Langley and Yalden, I feel it unlikely that pure bred wildcats should be still existing in Norfolk at such a recent date. As we have seen in other parts of Britain, however, it has been hypothesised that the remnants of the indigenous population has interbred with the burgeoning population of feral cats to produce a fearsome hybrid.

The taxonomy of the British wildcat is a puzzle, which we have discussed at greater length in the relevant chapter of *'The Smaller Mystery Carnivores of the Westcountry).*

There were numerous reports of porpoises this year from August onwards off Cley and Yarmouth. Most were of dead specimens, including two young ones, one three foot and the other two feet and four inches long, found with no external injuries on Yarmouth beach in early December. Other records were at Breydon on December 29th, one about three foot long and badly gashed on Hunstanton Beach on December 28th, and singly at Sea Paling in January, East Runton in March and Holkham in August.

MAMMAL REPORT FOR 1965

Again no colour variants in hedgehogs this year, but an interesting observation was made in Norwich, where a female gave birth to three young on July 17th and promptly ate them on the 18th.

There were two reports of mole variants. One, an albino with a ginger patch underneath, was caught at Thetford, and a silver-grey specimen at Binham. One mole at Horsford was seen swimming along the bottom of a stream, and another crossed a bridge at Bawburgh to avoid getting wet.

The hare was less abundant in 1965 than previously, but a black specimen was caught at Langham.

The melanistic water voles were still present in good numbers, and there were a few 'gingerish' specimens seen at Tasburgh and Whitlingham. In Wymondham, a short lived plague of the animals found their way into somebody's garden and fed on tulips and Sweet Williams, and stored 'great numbers' of Tulip bulbs in their burrows.

There were no reports of variant rats, but disturbingly reports from Didlington and Attlebridge suggested that rats were becoming immune to the rat poison Warfarin. This had also been reported in Montgomeryshire.

Possibly due to the recovery of the rabbit population stoats were also increasing in numbers.

This also resulted in there being more ermine animals than there had been for a while. There were several reports from Bawburgh, Surlingham, Strumpshaw, Didington, Great Snoring and Holt.

The mink was still surviving in Norfolk despite the belief that it may have been wiped out by the weather and hunters. One definite specimen was seen diving from an overhanging branch into the Yare at Colney. Another possible mink was recorded from the Wensum, well within the Norwich boundary.

There were more reports of feral domestic cats than there had been for some time. At Brooke they were described as "far too numerous". The "great number" at Horsford were mainly tabby in colouration. A population was reported living in railway suburbs at Caistor. Others are reported from Blakeney, Great Snoring, Attlebridge and Aylsham.

There was just one species of cetacean noted during 1965: the porpoise. Up to six at a time were seen off the North Norfolk coast. Two decomposing carcasses were found between Bacton and Mundesley on the 2nd December.

MAMMAL REPORT FOR 1966

Moles have been seen in high numbers this year with many more than usual seen above ground. There were several colour variants observed in North - chocolate brown, light brown with orange undersides, and a white specimen with a ginger belly which was kept alive for a week with some 250 worms a day, but which died, apparently from starvation! Six from the same area at Brinton had an oval patch of short fur on their backs.

The rabbit population is still fluctuating because of the myxamatosis disease and this has also taken a toll on the colour variants. There was just one melanistic rabbit this year, which was reported from Weston Longville. There was also a white specimen seen at Honningham, which was probably wild.

The black water voles have been reported from several areas. A five minute fight between a black and a brown specimen was watched at Skeyton. This resulted in the black creature winning.

There were few reports of red squirrels during the year, and even fewer melanistic specimens from anywhere other than Ellingham. However, a very dark, almost black specimen was seen at Horsford. There were no reports of grey squirrels.

This was not the most productive of years for unusual sightings and this trend continued with both mink and feral cats. A female melanistic mink was observed at Broome and cats were still numerous at Horsford.

Many small schools of porpoises were reported, one in conjunction with a basking shark on October 9th. Some were even seen between some swimmers and the shore. Two dolphins, tentatively identified as bottle nosed, were found dead during the year; one at Horsey in April and the other at Scott Head in July.

MAMMAL REPORT FOR 1967.

This was not a good year either for colour variations in Norfolk. There are no records of the more common animals, such as the mole or the rat, although there are a few others.

Some sandy coloured rabbits were observed at Tottington and black ones at Newton Flotman, although it is not known whether or not these animals may have been the result of wild rabbits interbreeding with escaped pets.

The house mouse is a common animal in Norfolk. Records were rarely reported, however, except for the occasional colour variant. The earliest such record was from 1967 when a melanistic mouse was found in Norwich City Centre.

The red squirrel is still holding on well in the county, despite the confirmation of grey squirrels at Ringland and Norwich City Centre. A melanistic specimen was seen at Horsford in July. Only one mink was recorded during the year - it was caught by a vermin trapper at Gillingham.

MAMMAL REPORT FOR 1968.

1968 welcomed four new species to the mammal report and there were several sightings of odd coloured animals - this despite the harsh weather which included one record of three inches of rain in a single night during September.

The mole was one of the species which suffered during the September floods and many starved to death from lack of food. Just one colour variant of this species was recorded during 1968 - a silver coloured animal seen at Thetford in August.

Three water shrews found dead in a tin were found to have an orange/russet underside, a black dorsal surface and white ear tufts.

A new species for the county arrived in the form of a parti-coloured bat which was discovered in a timber yard at Yarmouth Docks on August 28th. Its normal range is northern and central Europe and Asia but it has occasionally been seen in Britain. The timber came from the Baltics and the bat probably came with the timber.

Rabbits were seemingly unaffected by the harsh weather and several colour variations were seen during the year. On March 17th a melanistic specimen was seen at Hillington, a small colony of black ones were seen near Jubilee, Great Hockham; a black one at Stanford; a pale sandy coloured one was seen all summer at Foxley and an albino one was seen at Weeting on July 29th.

The melanistic water voles were still common in the north of the county and for the first time in many years they were reported from further south. Two were seen in a garden in Gorleston in July, and they were also reported from the River Thurne at How Hill. One was also seen in August on the Tas Marshes at Caistor.

'Locally abundant' was probably the best way to describe the house mouse during 1968. In some parts of Norwich they were found to be immune to Warfarin. Three white specimens were also seen in the city.

1968 was a good year for rats which are well represented throughout the county. Abnormal coloured specimens were reported from several localities. One, found dead on a road was a reddish-brown colour, while during August, a sandy-coloured one was observed sitting in a tree during a rain storm. A silver coloured rat was shot at South Creake in February.

The hind part of the skeleton of a black rat was discovered by archaeologists on the site of the Augustine Friary in Norwich. The rat was probably trapped when the Friary was demolished in the 16th Century. Fortunately, the species is now extinct in Norfolk, except where isolated specimens manage to get into the ports around the coast.

The chinese water deer was first introduced into the United Kingdom at Woburn Park in Bedfordshire in the early part of this century. Several escaped to Bedfordshire, Buckinghamshire, Hampshire, Northamptonshire and Shropshire. In 1968 two small deer were seen on several occasions at Swim Coots and Ling's Mill, Hickling.

It was thought that they were muntjac, but they proved to be chinese water deer when one was knocked down and killed on the Stalham by-pass. Both specimens were males.

Another new record for the Norfolk Mammal Report came during 1968 in the shape of the ferret. An albino male weighing two pounds was found dead on the Norwich ring road at hellesdon on February 12th. A polecat-ferret was recaptured at Thorpe late in the year after having been wild for some time.

The beech marten is another animal that is not indigenous to the United Kingdom. One which escaped from Great Wittingham Wildlife Park on August 26th 1967 was recaptured in May 1968 at Ringstead, 28 miles away. To quote from the Mammal Report:

"The escape and later destruction of this single somewhat inconspicuous animal demonstrates two things; firstly, any unusual animal found in Norfolk is highly likely to be an escapee, and secondly how effective the keepering of the region still is".

EDITOR'S NOTE: Whilst it is true that as Justin states, it is generally believed that the Beech Marten *(Martes foina)* is not a native to the United Kingdom, I have gathered information to suggest that this might be a zoological fallacy. My deliberations on the subject can be found in my book *"The Smaller Mystery Carnivores of the Westcountry"* (1996)

MAMMAL REPORT FOR 1969.

There was not much reported this year in terms of colour variants. Generally it was a bad year all round - the most disturbing news being the increase in the number of grey squirrels in the county. It was hoped that the fens to the west, the Breck pine forests and the Waveney would hold up the pests, but this was not to be.

The mole recovered well from the downpours of the previous year, and during 1969 was back on the increase. White, or partially white specimens were recorded from Cawston, Foulsham, Melton Constable and Kelling whilst what was described as a 'colony' of these animals was discovered at R.A.F Sculthorpe during the summer. A single cream coloured animal was found at Weybourne.

The rabbit population has either increased or decreased rapidly over the years as a result of the depredations of the myxamatosis virus and the resultant swings in population. It was during 1969 that it was found that mosquitoes in Norfolk were carriers of the virus. This was confirmed in October when insects caught at Wheatfen were examined at Monkswood Research Station.

Several colour variations amongst the wild rabbit population were reported during the year. Several black ones were seen at Horsford, a juvenile black specimen was seen in a hay field at Caistor St Edmunds on May 28th. a single black specimen was seen at North Elmham, twelve 'smokey blue' animals were shot during the year in the area from Stanford to Tottington, and three apricot coloured animals were seen at Little London.

Despite the abundance of the species few colour variations of the hare are reported from the county. In 1969 there was only one report. This animal was silver grey above and white beneath, the grey being almost uniform on the flanks and sides but streaked with darker hairs on the back, while the tail had a normal black patch on top.

The population of black water voles was increasing during 1969 in north Norfolk, where on the Glaven about ten percent of the population were of this colouration morph. One specimen was seen to climb about twelve feet up a hawthorn bush in order to eat the newly burst leaf buds.

There were two reports of melanistic red squirrels during 1969. An almost black specimen was observed at Walshingham on October 15th, and a black one was seen at North Elmham.

Another chinese water deer was discovered dead by the roadside on the A122 between Outwell and Nordelph on September 15th. It was a buck and weighed 43 pounds.

The ermine stoat is most common in the north of the country which tends to have colder winters. Four were reported from northern Norfolk, however, in late January and during February. A most exceptional report was received from Whitwell were a stoat was seen up a tree about 40 feet from the ground, being attacked by many small birds as it·ran along the branch.

MAMMAL REPORT FOR 1970.

Again this was a poor year for reports of either rare or oddly coloured animals. This is somewhat summed up by the recording of just one odd-coloured mole - an albino seen at Cranworth during February.

Several colour abberations of the common shrew were recorded in 1970. The most common of these were animals with white ear tufts. These were seen at Mousehold Heath along with a specimen that had eight white spots on its back.

EDITOR'S NOTE: This is by anyone's standards a striking, if zoologically inaccurate lexilink!

Three generally dark specimens were reported, whilst one reported on Stoke Marshes was a mixture of almost black with the more usual browny-coloured head. This animal also had the white ear tufts.

The most striking water shrew record came from Upper Sheringham, where a piebald specimen with a pelage similar to that of Dutch rabbits was picked up on May 23rd. The bold black and white striping started with a dark coloured nose and ended after six stripes with a pale stripe to the tail.

Colour variants in rabbits were also rare during 1969 with only one definitely wild black specimen being recorded from Horsford. Black and white ones apparently living wild were recorded at Wymondham and West Caistor.

Black coloured water voles were recorded from widely separate areas. Three to five were found at Corpusty and another specimen was found dead in a drainage pipe at Brinton. One of the Corpusty animals was seen eating pond snails on the branch of an Alder overhanging into the water.

The red squirrel appeared to be hanging on despite the increase in the gery squirrel population.

Another black specimen was reported from Horsford and an albino - the first ever reported from the county - was observed at East Runton in March.

Mink appeared to be enjoying life in the county during 1970, as they continued to live and breed succesfully. In Wroxham, one was discovered living in the roof of a thatched house. This specimen was eventually captured using fish as bait. During the year twelve were killed at Weiney Wildlife Reserve.

MAMMAL REPORT FOR 1971.

This was another quiet year for sightings of unusual animals. There were no odd animal records and very few of any colour variants of any significance. The main matter of concern was again the increasing number of grey squirrels in the county, which many observers saw as the beginning of the end for the indigenous species.

Two odd coloured moles were noted during the year. One trapped at Upton on March 18th was an orange/yellow colour. The other was a cream specimen seen at Welney in December.

The rabbit continued its remarkable population rise during the year, and they were increasing in numbers acros the county. At East Carleton a white individual was observed on October 25th in the same area as a white pheasant which had been seen the previous week. Black animals were seen at Whitlingham in August and 'two black or very dark' animals were seen in a group of seven animals seen just outside Norwich Railway Station.

A pure white hare was seen on several occasions in a field near Bixley during the early part of the year. This was described as a particularly strange spectacle when seen at dusk.

The water vole numbers were on the decrease over much of the county although the black individuals were still well represented. There were reports from Cley, Stifkey Marshes and around Watton. The most interesting report came from Corpusty where a family of nine invaded a garden. Two of these, (one of which was an adult) were largely brown, but with a well defined band of black guard hairs running down the back. Four had the black area extended down the sides turning to brown at a fairly sharp line giving a general appearance of being all black. The remaining three were almost all black with only the undersides being brownish-white.

No unusual red squirrels were sighted during 1971, but an unusual piece of behaviour was reported during the year at Duke's Bridge in Norwich. At about midday on June 16th, a squirrel was seen clambering along the top of a branch overhanging the river and did a 'high-board dive' with its feet spread out, landing in the river. After hitting the water, it began jerkily swimming across the remaining twenty feet to the other side, using front and back legs simultaneously with the tail held erect out of the water. Upon reaching the other side it promptly ran off.

The ferret continues to appear to surprised observers, and this year there were two reports. On June 4th one was found dead on the A11 near Bridgeham Heath, and an animal resembling a ferret was found on a gibbet at Hunstanton in October.

There had been few cetacean reports over the previous few years and this trend continued during 1971. Just two porpoises - one a dead juvenile - were seen in north Norfolk. On July 16th, however, a dead 35 foot lesser rorqual was discovered washed ashore. It had originally been seen off the Lincolnshire coast, and the next day had been seen five miles further south. The previous record had been in 1959.

MAMMAL REPORTS 1972-1974.

These years were again short on unusual sightings. The main highlight was a roe deer injured by a car at Thetford Chase. The back legs, shoulder, head and one ear were white, although the saddle was the normal colour.

MAMMAL REPORTS 1975-1993

Black rabbits were on the increase in 1975 and were recorded in good numbers throughout the county. The melanistic red squirrels at Ellingham Hall appeared to have died out along with the majority of the population in the County, although a very dark form was seen at Aldeby in September 1975 and a hardy red forager was seen in freezing temperatures at Cockley Cley in December the same year.

Perhaps the oddest report of 1977 was the water shrew which somehow found its way into a mouse trap in a cupboard at Mill-House, Corpusty. At Brooke, the same year, a golden mole was caught by a cat and later rescued.

Odd coloured moles have been recorded more frequently in the last few years. In May 1978 a cream/golden specimen was taken live at Stokesby, and in 1979 two golden specimens were observed at Dittinchingam. 1980 brought the capture of an albino pygmy shrew at Wissonsett in September, whils at Hockham, an all-cream hare was observed and seems to have produced one or two all-cream offspring as on two occasions two of this colour were seen together.

A most unusual discovery was made in 1981 when a hare was caught by dogs and was found to have been carrying four embryos in various sizes -1, 2, 4 and 6 inches long. The most exciting sighting of latter years was the walrus which appeared in the River Ouse during 1981. This incident is covered in more depth in *Animals & Men* #11 (October 1996).

The lack of information continued into 1982, the only item of interest being a sighting of a new colour variation for the rabbit - ginger and mahogany specimens reported from Hillborough and Buxton Heath.

1985 was a good year for bird watchers with several new species added to the county list. the most impressive of these was the black and white warbler which was seen at How Hill near Ludham. Well over two thousand birders, including myself, viewed the bird over a period of two days.

As already noted, odd coloured hares are rare in the county, and smokey-blue specimens especially so. One, however, was seen grazing marshes at Salthouse and at Leatheringsett during 1985. It seemed to lack the yellow pigment thus appearing silver-grey. darkest on its back and shading to almost white below. It had the usual dark tips to the ears and the tail and around the eye.

The strangest story of 1986 is that of the hare which was chased onto the sand dunes at Sea Palling by a dog and ran into the sea where it promptly drowned. However, the mammal of the year was undoubtedly the sperm whale which was washed up dead at Holkham.

Sperm whales had not been recorded in Norfolk for over two hundred years. They are inhabitants of almost all the world's oceans and seas north and south of the equator within the 40' meridian. The previous Norfolk reports had been in 1753 when a dozen were seen off the coat. These included two at King's Lynn.

On the weekend of 29th/30th November 1986, a dead specimen appeared on the beach at Holkham. It was identified as a Great Sperm Whale (Physeter macrocephalus) and within a few hours of its discovery was observed by many amazed spectators who touched it and in some cases clambered over it. It was a male and measured just over sixteen metres in length from the snout to the fork in the tail. It weighed an estimated 50 tonnes. Its general body colour was a dark, steely grey, but around the head were pale lines and blotches - the former are supposed to be injuries sustained by squid's tentacles as they try to escape from the mouth of the great beast.

The local council are obliged to dispose of unwanted health problems and this was no easy task as it took several days with the help of a mechanical digger and several skips. All of it was processed into oil and fertiliser except for the head, which being too heavy to move, was subsequently buried in the sand.

In 1991, just five years after the first sperm whale stranding, another turned up, this time at Scolt Head on the 12th November. Tides later washed it onto Brancaster Beach. The carcass was an estimated forty tonnes, and the problems of removing the creature were somewhat greater than before due to the advanced state of decomposition.

Two years later this 'surge' of whale beachings continued with a third specimen on December 2nd when a forty foot creature was stranded in the Wash off Beacham. It died soon after stranding in the early morning tide and was washed out to sea the following evening.

To date there have been no more sperm whale strandings off Norfolk. It is not known why they have appeared in the last few years. Perhaps the search for food drove them ever closer to the shore and they were caught out in the many sand dunes present in the area. Perhaps, as the available prey species decline more will follow, and of other whale species as well!

There are two other cetacean records of interest. In 1991 a school of pilot whales was observed off the coast at Holme. A single specimen seen the same January off Tichwell was listed as a possible killer whale.

We go back to terrestrial mammals for the final report for this paper. An odd coloured fox was seen in 1993 between Snettisham and Immere. It was a partial albino with a white head and pink eyes, a normal body and a white tail. Unfortunately this peculiar beast was never seen again!

EDITOR'S NOTE: Justin also included this following item, which although not of a mammal, is certainly of interest to the fortean zoologist:

GEESE KILLED BY ELECTRICAL STORM - 1978

The following is an account from the eastern Daily Press concerning the fierce storm that tore through Norfolk on 12th January 1978:

"East Anglia experienced a storm of exceptional violence when the advance of a cold front coincided with the very sharp fall of the barometer. In addition to the impact of severe gales, with whirlwinds causing havoc in some places, almost the whole of Norfolk was enveloped by a downpour of sleet, hail and snow accompanied by lightning and thunder virtually unprecedented in winter.

At the onset of the gale most birds took cover or became grounded but the effect on wild geese in the vicinity of the Wash was to cause general so that the flocks arose and fled. But they were caught up in the turbulence of whirling hail and a scene so charged with electric violence that it had the fury of a full scale battle. In the space of an hour, no less than 136 geese were killed in flight as they travelled downwind with the storm. They dropped dead out of the sky along a track some eighteen miles in a direction somewhat south of east. More than a hundred fell on the farm at Castleacre, fifty seven being found piled into a heap at one spot.

At some points along the route the victims were found singly but at others ten or more were discovered in line over short distances. In many cases the geese showed no sign of external injury and no scorching of their plumage had been reported, but in many cases narrow strips of feathers had been lost from the wings and breast.

There can be little doubt that these birds were killed by lightning, though rather from its explosive action than by direct electric shock. A similar event, though not on the same scale, took place in north Norfolk in February 1905. On that occasion fifteen pink-footed geese and *four white-fronted geese were reported killed by lightning when a fierce snow storm blew up suddenly from the west".*

EDITOR'S NOTE: justin Boote should be commended for this mammoth piece of research, and I hope that it acts as a benchmark for those who follow. It is interesting to see how closely the sites of records of colour variants mirror the sites of records of both unknown animals and less tangible parapsychological phenomena.

This is a fascinating ongoing project, and one which I aam looking forward to bringing to fruition.

* * * * *

According to Louis Baba

by Roy Kerridge

I have a penfriend in Ghana called Louis Baba. He is a university educated snake and animal catcher for the zoo and pet shop trade, and has been bitten by poisonous snakes many times but has survived. His letters shed unusual light on the Dark Continent. Here are some examples:

"The people by the roadside, noticing my cage, began to talk about my snake handling activities. This rattled one woman, so that she said, very loudly, 'if he were my husband, my God I wouldn't marry him!'

We were stranded for some hours waiting for a truck to take us to town. One man told me about a time that he had been stranded for days. When at last a truck for carting cocoa came along, all the people had crammed themselves in, leaving no space for the slight movement of bodies. The driver over filled his pocket with the fares that he had collected, and took off. The road, seldom used, was bad, strewn with bamboos laid low by the assaults of the wind. The driver rammed on clanking through them. Just when they had cleared the bamboo, a gigantic green mamba snake had fallen on top of the passengers. It had been basking leisurely on a stalk.

Many people, on seeing a green mamba, which we call 'Ochiribin', wanted to jump off, preferring to sustain broken bones than stay and get bitten by a deadly snake, but the high body of the truck and the cramped conditions made jumping off impossible. Senseless yells were produced. The green mamba had realised where it was, and searched frantically for a way out.

'Come and see', as the teller related, 'Ochiribin riding triumphantly on people's shoulders!'

It travelled to and fro several times, trying to settle its head in the bridge of people's noses, though they tried to veer their heads away. Nevertheless, Ochiribin went on riding. In the commotion, many people complained that they had been bitten. At last, the snake found a heap of empty sacks in the corner, and slipped under them. The driver stooped and there was time to examine those who had been bitten. It was found that all claims had been imaginary.

Thereby a referendum ensued, and it was decided that as the stranger had done no harm, it be let go in peace. As soon as the boldest man had lifted the sacks, Ochiribin, as of aware of what had transpired, took its leave and slipped down calmly onto the road, slowly entering the nearby bush. So one again, the awesome reputation of the green mamba has been brought to question. The passengers cautioned the driver not to ram through bamboos again. They needn't have bothered - such good events never come twice a day".

AHDCBX Alamy Images

Green mamba

Trouble on the Ghana-Upper Volta border -

"Borders of both countries are closed, but the natives on both sides walk to and fro each day. There is an unguarded stream that marks the boundary.

A Ghanaian border guard soldier once pursued a smuggler over into the Upper Volta side. Natives came out with bows and arrows and fell upon the soldier, tore his gun and uniform from him and carried him to the local chief. How they said the episode ended must have been very nasty for any soldier. We have the most warlike tribe around in these parts; Moshes.

Gypsies coming here from the Sahara desert have from time to time fed my curiosity. One fair evening I acted as interpreter between a small Ashanti lad and a gaunt looking, Hausa-speaking gypsy woman. A wisp of a woman, she was peddling potions for subduing heavyweights and smashing curses. The Ashanti lad wanted charms for winning the love of any woman he felt for. Next he wanted powers by which he could knock down decisively any man who might try and halt his love winning spree. A ring was said to ensure that. The transaction was topped off with a purchase of aphrodisiacs.

There is a white American on a visit here, anxiously seeking to acquire supernatural powers. Quite strange for me, to see someone coming thousands of miles out of a velvet garden to seek hidden ingredients of the grey woods. His contact man took him to the northern regions, where, I suppose, for a handsome fee, amulets were sewn for him to wear over his biceps. Back in his Acorn Hotel, he is very happy. He has befriended the porter of the hotel who has tribal marks on his face, and they intend to travel back together to the U.S. There the American has four bodyguards, and intends to make the porter the fifth.

I heard the American ask the porter, 'Do you know juju?' For a reply the porter pulled up a sleeve to reveal amulets and went on to say, 'I will take juju to America, and when they shoot at me, I will vanish'. Though the potency of the amulets remains untested, he is too confident to worry. So sure is he of vanishing at the sound of a gun, that at the pronouncing of the word 'vanish' he glees. He says that when they have all shot at him in vain, he will finally return to Ghana with a 'portmanteau full of American dollars'. What to do with so many dollars he doesn't say. The absurdities aside, there is a danger he isn't taking into account. Why bank so much on being shot at unless he is being introduced to outlaw games?"

With that, Louis Baba signed off, "Yours sincerely", leaving me to ponder on the fate of the porter, and the identity of the American.

Most likely, the American is a pop singer and drug addict, whose trigger-happy enemies exist only in his paranoic imagination. In that case, the porter will come home safely with his portmanteau, and live happily ever after. To find out more I will have to await the next missive from that master letter writer, Louis Baba.

1996:

A Year in the life of the Centre for Fortean Zoology.

To misquote Oscar Wilde. "To lose one cartoonist is unfortunate, to lose two looks like carelessness". To lose our founding cartoonist Jane Bradley in 1995 was bad enough, but during 1996 we lost her replacement, and also three other members of the core team of the CFZ. We gained several invaluable new recruits, and although this is a year which Her Majesty would probably have described as an 'annus horribilis', it wasn't all bad and we managed to achieve quite a bit.

The first disaster of the year took place in the early spring when Martin Brown, the talented cartoonist who was known to readers of *Animals & Men* as 'Mort' killed himself. The second took place in July, when Alison, my wife and co-founder of the CFZ left me, and took with her my quasi daughter Lisa, who had been responsible for much of the A&M Artwork and demanded a divorce. In the ensuing unpleasantness we also lost John Jacques, our psychotherapist and interviewer.

After having had two cartoonists die on us we felt very wary about offering the job to anyone else. When Mark North telephoned us and offered to take over both Lisa's and Mort's roles we were overjoyed. We warned him about the fates of his two predecessors but he said something about 'third time lucky', and has been delighting us with his surreal view of the omniverse ever since.

The day that Alison left me I telephoned Graham Inglis. He has been a friend of mine for nearly a decade now, and has worked for me in various capacities, usually involving carrying heavy items of equipment for very little remuneration. He has impeccable green credentials, and is somewhat of an administrative genius (having worked as an administrator for the DOT), and was prepared to work for nothing. He also wanted the opportunity to write about Xenobiology and debunk what he described as 'psychic piffle'. He was the obvious person for the job.

During this turbulent year various people who had previously been of minor importance within the hierarchy of the CFZ suddenly stepped into the centre stage. Tom Anderson, a bizarre Celt, who dislikes my appelation of him as *'Aberdeen's Mr Entertainment'* had been writing for us for ages but it was with the summer of 1996 and its attendant upheavals that he really came into his own. Darren Naish should also be thanked for his prodigous input of articles and pictures, as should Clinton Keeling and Chris Moiser. But they are just four of many!

Despite all the problems we still managed to produce three books, four issues of *Animals & Men*, a reprint of the 1976 booklet about the Falmouth Bay Sea Monster and a highly successful cryptozoological exhibition for the Unconvention in April.

At the end of the year we launched a second magazine, *'The Goblin Universe'*, and despite technical hitches with the printer (see the introduction) we still managed to enter what is euphemistically known as the festive season with our heads held high and our fortean credibility relatively intact. (And, oh yes, we managed to solve the riddle of the sea monster carcase of Durgan Beach).

The future may not be exactly what we had planned twelve months ago, but it still looks relatively bright. Although further technical hitches meant that neither this yearbook or the 'owlman' book actually saw the light of day until 1997, we are now (I hope) pretty well back on course.

Jon Downes.
Exeter,
March 1997.

THE CENTRE FOR FORTEAN ZOOLOGY

So, what is the Centre for Fortean Zoology?

We are a non profit-making organisation founded in 1992 with the aim of being a clearing house for information, and coordinating research into mystery animals around the world. We also study out of place animals, rare and aberrant animal behaviour, and Zooform Phenomena; little-understood "things" that appear to be animals, but which are in fact nothing of the sort, and not even alive (at least in the way we understand the term).

Why should I join the Centre for Fortean Zoology?

Not only are we the biggest organisation of our type in the world, but - or so we like to think - we are the best. We are certainly the only truly global Cryptozoological research organisation, and we carry out our investigations using a strictly scientific set of guidelines. We are expanding all the time and looking to recruit new members to help us in our research into mysterious animals and strange creatures across the globe. Why should you join us? Because, if you are genuinely interested in trying to solve the last great mysteries of Mother Nature, there is nobody better than us with whom to do it.

What do I get if I join the Centre for Fortean Zoology?

For £12 a year, you get a four-issue subscription to our journal *Animals & Men*. Each issue contains 60 pages packed with news, articles, letters, research papers, field reports, and even a gossip column! The magazine is A5 in format with a full colour cover. You also have access to one of the world's largest collections of resource material dealing with cryptozoology and allied disciplines, and people from the CFZ membership regularly take part in fieldwork and expeditions around the world.

How is the Centre for Fortean Zoology organized?

The CFZ is managed by a three-man board of trustees, with a non-profit making trust registered with HM Government Stamp Office. The board of trustees is supported by a Permanent Directorate of full and part-time staff, and advised by a Consultancy Board of specialists - many of whom who are world-renowned experts in their particular field. We have regional representatives across the UK, the USA, and many other parts of the world, and are affiliated with other organisations whose aims and protocols mirror our own.

I am new to the subject, and although I am interested I have little practical knowledge. I don't want to feel out of my depth. What should I do?

Don't worry. We were *all* beginners once. You'll find that the people at the CFZ are friendly and approachable. We have a thriving forum on the website which is the hub of an ever-growing electronic community. You will soon find your feet. Many members of the CFZ Permanent Directorate started off as ordinary members, and now work full-time chasing monsters around the world.

I have an idea for a project which isn't on your website. What do I do?

Write to us, e-mail us, or telephone us. The list of future projects on the website is not exhaustive. If you have a good idea for an investigation, please tell us. We may well be able to help.

How do I go on an expedition?

We are always looking for volunteers to join us. If you see a project that interests you, do not hesitate to get in touch with us. Under certain circumstances we can help provide funding for your trip. If you look on the future projects section of the website, you can see some of the projects that we have pencilled in for the next few years.

In 2003 and 2004 we sent three-man expeditions to Sumatra looking for Orang-Pendek - a semi legendary bipedal ape. The same three went to Mongolia in 2005. All three members started off merely subscribers to the CFZ magazine.

Next time it could be you!

Project Kerinci, Sumatra - 2003
In search of the bipedal ape Orang Pendek

How is the Centre for Fortean Zoology funded?

We have no magic sources of income. All our funds come from donations, membership fees, works that we do for TV, radio or magazines, and sales of our publications and merchandise. We are always looking for corporate sponsorship, and other sources of revenue. If you have any ideas for fund-raising please let us know. However, unlike other cryptozoological organisations in the past, we do not live in an intellectual ivory tower. We are not afraid to get our hands dirty, and furthermore we are not one of those organisations where the membership have to raise money so that a privileged few can go on expensive foreign trips. Our research teams both in the UK and abroad, consist of a mixture of experienced and inexperienced personnel. We are truly a community, and work on the premise that the benefits of CFZ membership are open to all.

What do you do with the data you gather from your investigations and expeditions?

Reports of our investigations are published on our website as soon as they are available. Preliminary reports are posted within days of the project finishing.

Each year we publish a 200 page yearbook containing research papers and expedition reports too long to be printed in the journal. We freely circulate our information to anybody who asks for it.

Is the CFZ community purely an electronic one?

No. Each year since 2000 we have held our annual convention - the *Weird Weekend* - in Exeter. It is three days of lectures, workshops, and excursions. But most importantly it is a chance for members of the CFZ to meet each other, and to talk with the members of the permanent directorate in a relaxed and informal setting and preferably with a pint of beer in one hand. Since 2006 - the *Weird Weekend* has been bigger and better and held in the idyllic rural location of Woolsery in North Devon. The 2008 event will be held over the weekend 15-17 August.

Since relocating to North Devon in 2005 we have become ever more closely involved with other community organisations, and we hope that this trend will continue. We also work closely with Police Forces across the UK as consultants for animal mutilation cases, and we intend to forge closer links with the coastguard and other community services. We want to work closely with those who regularly travel into the Bristol Channel, so that if the recent trend of exotic animal visitors to our coastal waters continues, we can be out there as soon as possible.

We are building a Visitor's Centre in rural North Devon. This will not be open to the general public, but will provide a museum, a library and an educational resource for our members (currently over 400) across the globe. We are also planning a youth organisation which will involve children and young people in our activities. We work closely with *Tropiquaria* - a small zoo in north Somerset, and have several exciting conservation projects planned.

Apart from having been the only Fortean Zoological organisation in the world to have consistently published material on all aspects of the subject for over a decade, we have achieved the following concrete results:

* Disproved the myth relating to the headless so-called sea-serpent carcass of Durgan beach in Cornwall 1975
* Disproved the story of the 1988 puma skull of Lustleigh Cleave
* Carried out the only in-depth research ever into the mythos of the Cornish Owlman
* Made the first records of a tropical species of lamprey
* Made the first records of a luminous cave gnat larva in Thailand.
* Discovered a possible new species of British mammal - the beech marten.
* In 1994-6 carried out the first archival fortean zoological survey of Hong Kong.
* In the year 2000, CFZ theories where confirmed when an entirely new species of lizard was found resident in Britain.
* Identified the monster of Martin Mere in Lancashire as a giant wels catfish
* Expanded the known range of Armitage's skink in the Gambia by 80%
* Obtained photographic evidence of the remains of Europe's largest known pike
* Carried out the first ever in-depth study of the *ninki-nanka*
* Carried out the first attempt to breed Puerto Rican cave snails in captivity
* Were the first European explorers to visit the `lost valley` in Sumatra
* Published the first ever evidence for a new tribe of pygmies in Guyana
* Published the first evidence for a new species of caiman in Guyana

Other books available from
CFZ PRESS

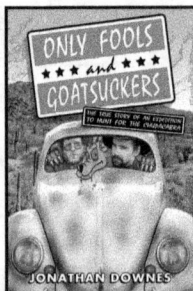

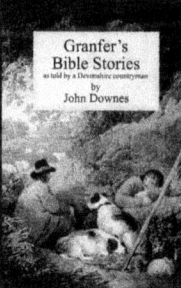

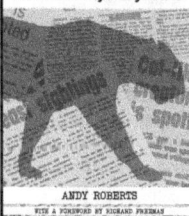

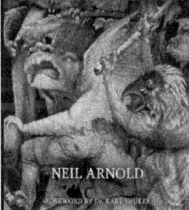

Other books available from
CFZ PRESS

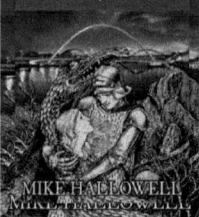